Toxic Torts Deskbook

M. Stuart Madden

CRC Press
Taylor & Francis Group
Boca Raton London New York

CRC Press is an imprint of the
Taylor & Francis Group, an **informa** business

First published 1992 by CRC Press
Taylor & Francis Group
6000 Broken Sound Parkway NW, Suite 300
Boca Raton, FL 33487-2742

Reissued 2018 by CRC Press

Library of Congress Cataloging-in-Publication Data

Madden, M. Stuart, 1948-
Toxic torts deskbook / M. Stuart Madden.
p. cm.
Includes bibliographical references and index.
ISBN 0-87371-508-X
1. Toxic torts--United States. I. Title.
KF1299.H39M33 1992
346.7303--dc20
[347.3063] 91-48238

A Library of Congress record exists under LC control number: 91048238

ISBN 13: 978-1-315-89825-4 (hbk)
ISBN 13: 978-1-351-07735-4 (ebk)

Visit the Taylor & Francis Web site at http://www.taylorandfrancis.com and the
CRC Press Web site at http://www.crcpress.com

To Mary-Anne O'Donnell Madden,
for her love and support,
and for Christopher and Michael.

Preface

The author and the publisher conceived of this book as a valuable resource for two distinct, but allied, groups of professionals. *Toxic Torts Deskbook* is written as a succinct, readable guide for professionals — attorneys and non-attorneys alike — who are affected by potential tort liability of those engaged in the manufacture, sale, marketing, or handling of chemicals and other potentially hazardous substances.

The text is intended to serve a dual purpose. First, it is a guide for all professionals to the toxic tort implications of the manufacture, marketing, and disposal of potentially hazardous substances. Second, *Toxic Torts Deskbook* serves as an initial reference for attorneys called upon to advise clients producing or selling such products or to legal counsel asked to consider present or potential legal claims from either a defendant's or a plaintiff's perspective.

The text should be of special value in aiding communication between professionals and attorneys. The chemical or industrial engineer, the manufacturer's corporate officer, or the professional who may be asked to serve as an expert witness in tort litigation will find this text quite valuable as a guide to the policy, rules of liability and defense, and the practical requirements of lodging a toxic tort claim or defense. If the text serves no purpose other than to help industrial sector professionals and legal counsel understand one another, the author will be satisfied.

To succeed as a reference for professionals ***and*** for attorneys, I adopt a style of narrative and of citation to authority that is not conventional to legal writing. In terms of narrative style, I have tried to avoid assuming that all readers would be equally familiar with various Latin or shorthand legal terms. Therefore, for concepts ranging from negligence per se to "state-of-the-art" evidence, I have paused in my discussion of court decisions to state briefly and plainly the meaning, and the practical implications, of such theories. The purpose, initially, of this approach was to give non-attorneys a sufficient understanding of certain legal terms and doctrines so that the discussion to follow would deliver the greatest possible proportion of useful information. As preparation of the book progressed, I came to appreciate that many attorneys would also profit from a "plain English" description of many toxic tort concepts.

Regarding references to authority, for brevity, I have avoided a bottom-heavy citation approach and often employ only limited citations. For example, in describing a cause of action for engaging in "abnormally dangerous activities", or for breach of the "implied warranty of merchantability", an initial citation may guide the reader to a provision of the Restatement, Second, of Torts, or the Uniform Commercial Code, but omit continued and more particular citation to subsections or phrases discussed thereafter. A similar approach is taken to statutory references, such as, for example, the Federal Insecticide, Fungicide and Rodenticide Act or the Federal Hazardous Substance Act. Where I reference these statutes or others in the text, ordinarily I refer the reader to the United States Code citation only at the outset of any pertinent discussion. The text will, as appropriate, inform the reader from where the author quotes or paraphrases statutes or decisions. Following an initial citation, however, I will not invariably provide further detailed references.

To keep the text at a length optimal for ready reference to busy professionals, the author does not analyze the range of toxic tort issues that might appear in longer, even multivolume texts sold principally to attorneys. Methodical discussion is not devoted to, for example, the theory and application of comparative fault, or innocent misrepresentation. The author and the publisher believe, however, that a work of this scope describes effectively the metes, bounds, and application of the principal issues in toxic tort litigation. It should find valuable application in discussion between professionals and attorneys regarding the prosecution, defense, and avoidance of toxic tort litigation.

M. Stuart Madden
Professor of Law
Pace University School of Law
White Plains, New York

Acknowledgments

The author expresses his gratitude to Pace University School of Law for its institutional and financial support of this project. More specific thanks are due Dean Steven Goldberg; Nicholas Triffin, Professor of Law and Law Librarian; and the School of Law library staff.

My research assistant David Klein (J.D. 1992) provided valuable research, drafting, and editorial assistance. I appreciate as well the capable support of School of Law secretary Dolores Vouhnout.

Contents

M. Stuart Madden is a Professor of Law at Pace University School of Law, White Plains, New York, where he has taught courses in torts, products liability, and evidence. He received his J.D. in 1976 from Georgetown University Law Center, where he was an Associate Editor of the *Georgetown Law Journal*. From 1976 through 1985, the author practiced law, as an associate and then as a partner, in District of Columbia law firms. Professor Madden publishes extensively on torts subjects and is Editor-in-Chief of the *Journal of Products Liability*. He is author of the two-volume *PRODUCTS LIABILITY* (2d) (1988). His work is cited widely in court decisions, law review articles, and law school casebooks. The author contributes actively to the torts work of the American Bar Association — Torts and Insurance Practice Section and the American Law Institute.

CHAPTER 1

Introduction

CONTENTS

Introduction

1.1 GENERALLY

Toxic torts comprise harms to persons,[1] to property,[2] or to the environment[3] due to the toxicity of a product, a substance, or a process. In many circumstances, the toxic harm is latent for a period of time and is not discernible for years, or even decades. As a modern legal term, "tort", derived from the Latin term for "twisted", is a civil, noncontractual wrong for which an injured person may seek a "remedy" in the form of money damages. A tort is a *civil* wrong in the sense that the injured party's remedy is a civil suit for compensatory damages, as distinct from a suit brought by the government for civil or criminal money penalties.

Tort law, therefore, is the body of related doctrines imposing civil liability, in money damages, upon persons or businesses whose substandard conduct causes personal physical or emotional injury to others or damage to their property. An older definition of tort law as comprising liability for civil, noncontractual, nonstatutory harm is, today, somewhat misleading. For example, the remedy for breach of the implied warranty of merchantability, providing money damages for toxic harm, is, by its terms, contractual, but it is also quite tort-like in the proof required and the damages available to the successful litigant.

In addition, modern tort law is substantially interwoven with provisions of state and federal statutes pertaining to such subjects as burdens of proof, comparative fault, and statutes of limitation, to name only three. By way of illustration: (1) a business's failure to comply with a standard of care established by a labeling regulation might, under state law, be deemed negligence per se (presumptively negligent); (2) a state or federal environmental statute may vest in individual citizens the right to pursue private cost recovery lawsuits, called "citizen suits", against a polluter; or (3) a state products liability statute may provide criteria for evaluation of issues ranging from the hazards posed by a product or substance to the varying burden of proof plaintiff must sustain against different participants in the chain of manufacture, marketing, and distribution.

Even with the contemporary influence of statutes and warranty law, the goals of tort law remain these: (1) assignment of responsibility, in money damages, to those responsible for creating a risk that produces harm; (2) compensation of persons for loss caused by another's substandard conduct; (3) deterrence of further unreasonably hazardous conduct by the responsible party and others engaged in similar pursuits; and (4) encouragement of innovation, such as changes in design, formulation, packaging, labeling, transportation, disposal or the like, that will reduce or eliminate unreasonable hazards.

In a toxic tort suit, plaintiff sues the responsible party for compensatory damages, i.e., a monetary award calculated to remedy harm the defendant has caused, to the imperfect extent financial relief can. By "compensatory damages" is meant money damages to compensate the injured party for the loss or harm suffered. Where the toxic harm is to property, or to an ongoing business, "compensatory damages" can be assessed on the basis of diminution in value to a property, or "downtime", economic loss, or cleanup costs to a business. In these instances, compensatory damages may come close to placing the injured party in the economic position they would have enjoyed had they not suffered injury or loss due to defendant's conduct.

Where, on the other hand, the harm suffered by defendant's wrongdoing is personal physical injury, disease, or death, these physical and emotional harms cannot be truly "compensated" for, as no amount of money damages can remedy such a loss. Nonetheless, a judge's or a jury's finding of defendant's liability for civil money damages in a toxic tort claim provides the best recompense available through the U.S. legal system.

There are, of course, variations upon these general observations. Plaintiff may bring a noncompensatory, but remedial, "citizen suit" claim to force compliance with state or federal environmental statutes. In addition, where defendant's injurious conduct is continuing in nature, a court may enjoin (prohibit) further similar conduct. A defendant failing to comply with the requirements of such an injunction may be subject to civil or criminal fines.

Another and significant variation is that, in addition to compensatory damages, punitive damages may be awarded upon plaintiff's showing that defendant's conduct was of such a reprehensible nature as to warrant financial punishment in addition to the burden of compensating plaintiff for the injury sustained.[4] In instances where defendant's conduct is of a particularly culpable nature, such as where the evidence shows that he or she acted with knowledge of the likelihood that toxic harm would result, or with reckless indifference to the risks posed to others, plaintiff may also be awarded punitive damages. Punitive damages are not intended to compensate plaintiff for the harm suffered. Rather, the goal of punitive damages is to punish a particularly blameworthy defendant and to serve to deter that defendant, and others, from engaging in future similar hazardous conduct.

An injury may be personal, physical injury to a worker, a patient, or a passerby. It may be damage to property, its breadth ranging from rendering an individual premises uninhabitable to the contamination of a large watershed.[5] A person's physical injury may range from nausea[6] to neurological damage or death. Some toxic tort claims for personal physical injury involve physical injury not directly related to the toxic qualities of a substance but rather caused by an individual's abreaction[7] or allergic reaction thereto.

The toxic tort harm may arise from a worker's exposure to hazardous chemicals in the workplace[8] or her inhalation of respirable carcinogens.

Contamination of groundwater by waste disposal,[9] by inadequately treated industrial effluent, or by administration of residential or agricultural pesticides, may also create toxic tort liability.

Various health procedures may give rise to toxic tort claims. Such recent claims range from harmful radiological exposure to X-rays[10] to employment of dental surgery products containing paraformaldehyde that plaintiff claimed had leached into his blood system.[11] Health care products liability, most frequently associated with pharmaceuticals, is usually considered a matter of products liability and not toxic torts. Even so, court decisions in pharamaceutical products liability cases often have a bearing on the law of toxic torts. For example, suits against drug manufacturers usually require a judge to evaluate expert testimony and the qualifications of the experts giving that testimony. Questions often arise concerning the relationship between the expert's training and professional experience and the basis for the expert's conclusions as to the cause of the claimed injury.

Another example of the relationship between toxic torts and pharmaceutical products liability is the developing law permitting, or denying, judgment for plaintiffs claiming long latency injury where plaintiff is unable to identify a particular manufacturer's product, substance, or pharmaceutical as having caused the injury. Some courts have permitted recovery of damages for such injury on a theory of "market-share" liability, or related logic. In both settings, court decisions involving pharmaceutical products liability can guide professionals, counsel, and judges considering toxic tort questions. To this extent, the text will discuss drug products liability suits bearing on these and similar subjects.

Property damage due to toxic exposure may, in turn, be temporary and remediable or permanent. An example of the former would be some forms of groundwater contamination; an example of the latter might be a public building constructed with building products containing respirable asbestos, where the cost of safely removing the asbestos would be greater than the value of the building itself. Plaintiffs in a toxic tort claim may be private persons, suing individually or as a "class" of individuals suffering the same type of harm. Plaintiff may also be a business or corporation, or even a government entity, such as a city, town, or county, bringing suit for pollution or contamination of public property.

1.2 SPECIAL CHARACTERISTICS

1.2.1 Generally

A toxic tort claimant frequently phrases a complaint in terms of multiple theories of recovery.[12] In the wellwater contamination suit of *Merry v. Westinghouse Electric Co.*,[13] for example, plaintiffs brought suit under theories of negligence, strict liability for abnormally dangerous activities, trespass, and nuisance. Irrespective of the theory or theories of plaintiff's toxic tort claim, conventional tort rules of duty and compensable damages pervade the law of toxic torts, as do the newer products liability doctrines of causation and identification of unreasonable risk. For example, even in a modern personal injury action claiming a manufacturer of asbestos products failed to warn adequately, a court will readily turn to the common law negligence standard to find the applicable duty.[14]

1.2.2 Long Latency

Nonstatutory civil claims arising from toxic harms, or toxic torts, often involve personal physical injury or property that remains undiscovered for years after the injurious exposure. A shipyard worker's injurious exposure to respirable asbestos fibers may result in asbestos-related disease only years later. An electroplating plant's contamination of its property, surrounding property, or subterranean aquifers may only be discovered when, years after the injurious activities, a successor owner of the property wishes to sell it. The Viet Nam veteran or the agricultural worker exposed to dangerous levels of a chemical herbicide may only be diagnosed with neurological disease or other illness many years thereafter.

Because toxic tort claims almost always involve injury or damage that has a long latency period before the harm manifests itself, these qualities of toxic torts distinguish them from many other common law claims such as ones in negligence, nuisance, or products liability. Also, and frequently because of the long latency between exposure to the toxin and claimant's ability to recognize the harm, toxic torts practically always involve complex questions of medical or environmental causation. As a consequence, toxic tort litigation will always require the engagement of experts in medicine or other science to help the attorney, and the fact finder, appreciate the presence or absence of a causal relationship between the toxin and the harm.

1.2.3 Proof of Causation

In claims for toxic harm, plaintiff must demonstrate that defendant's product or activity was a direct cause of the resulting personal injury or damage to property. Whether the suit is brought in negligence, strict products liability, or warranty, defendant's actions must be shown to be the "proximate cause" of the harm. By "proximate cause" is meant the challenged act was (1) a substantial contributing factor in bringing about the injury and (2) the relationship between defendant's act and the injury or damage is not so remote or attenuated as to suggest that the harm was not foreseeable, i.e., that it would be neither fair nor reasonable to hold defendant responsible.

Conventional tort and toxic tort claims alike often raise complex issues of proximate cause. In a typical complex products liability suit not involving a toxic product, such as a "second collision" automobile design claim, plaintiff would assert that the vehicle posed an unreasonable risk of injury to its occupant during a collision. Plaintiff's quite difficult burden in such a suit is to prove that the design and construction of the vehicle was such that it exacerbated the injuries plaintiff would have suffered from the same collision in an automobile of a different construction.

Because years often pass from the injurious exposure to the manifestation of personal injury or property damage or contamination, the majority of toxic tort claims pose additional and distinctive problems in the proof of proximate cause. As a consequence, it is often quite difficult for plaintiff to identify the particular substance involved and the specific actor who was responsible. Years after exposure to respirable asbestos, a shipyard pipefitter may be unable to identify the particular manufacturer of asbestos products to which he was exposed. Even where a particular manufacturer's asbestos products, or chemical solvents, or pesticide can be identified, the passage of time may make it difficult for plaintiff to demonstrate the times of exposure, the duration of the exposure, or the level of the toxic substance to which he was exposed.

1.3 ROLE OF EXPERT SCIENTIFIC OR MEDICAL TESTIMONY

Proof of toxic tort causation may be more difficult than proof in many conventional personal injury claims. Where, for example, the disease suffered by a toxic tort litigant is cancer of the liver, plaintiff must demonstrate by a preponderance of the evidence, i.e., it is more likely than not, that the disease was caused by exposure to a product or a process for which defendant was responsible.

Proof of proximate cause in a toxic tort claim almost invariably requires that plaintiff introduce expert scientific or medical evidence that defendant's product or process was a substantial contributing factor in plaintiff's injury or loss. Indeed, the rule followed in most jurisdictions is that plaintiff cannot succeed in a claim for toxic personal physical injury without presentation of expert medical or scientific evidence. An expert witness, such as, for example, a physician testifying on plaintiff's behalf, will ideally testify that, to a reasonable degree of medical or scientific certainty, defendant's substance caused plaintiff's injury or disease. An epidemiologist testifying on behalf of plaintiff will, if supportable, testify that in plaintiff's particular case, it is more probable than not that defendant's toxic substance caused the injury. Defendant, in turn, will ordinarily counter with experts of its own choosing who have reached conflicting conclusions or who seek to discredit the diagnostic or statistical bases upon which plaintiff's witnesses relied. It is seen readily that most toxic tort claims involved quite complicated, laborious, and expensive litigation preparation.

In some contexts, vigorous scientific debate exists as to whether the toxic substance can, in any circumstance, cause the injury or disease that plaintiff suffers. As often, and even where defendant must concede that a certain level of exposure to a substance can cause injury, plaintiff must prove that the exposure to defendant's product was the producing cause of the injury. Defendant may be expected to produce evidence that the injury or disease was caused by (1) exposure to the product of another producer; (2) exposure to a different product or substance altogether; (3) exposure to background levels of toxins in the environment that affect the health of all persons in industrialized areas, rather than exposure to the toxin produced or released by anyone in particular; or (4) exposure to altogether different substances, ranging from tobacco products to alcohol, for which defendant should bear no responsibility.

A frequent quality of toxic tort claims is their scientific or medical novelty. Although the injuries themselves may not be new, e.g., the risks of disease from working in asbestos mining, which was recognized during the 1930s,[15] the cause-effect relationship of plaintiff's claim of disease is, nevertheless, almost always disputed vigorously by the parties. The manufacturer or other defendant will argue, where appropriate, that (1) the toxin could not be injurious to plaintiff on the facts of this case — e.g., defendant might argue that plaintiff's exposure to the substance was too remote (i.e., plaintiff worked at a work station significantly removed from the industrial use of the chemical in question); or (2) that the form of the toxin was such as to make exposure to it nonharmful (i.e., the asbestos-containing pipe fittings in question were resin-bonded, minimizing the release of respirable asbestos fibers).

Defendant might argue that the substance does not, in the context of the usage or exposure claimed by plaintiff, create an unreasonable risk of disease or injury, i.e., the position often taken by defendants in suits alleging disease from use of tobacco products. Lastly, defendant might argue that while the toxin or product might cause injury or disease to one in plaintiff's circumstances, it was not a cause of plaintiff's injury.[16]

In all of these settings, both plaintiff and defendant will need to produce expert witnesses to confirm their allegations or defenses. Such expert testimony is needed because jurors[17] are unable to make informed decisions as to the risk posed by most potentially toxic products, the relationship between plaintiff's exposure and the ensuing disease or loss, and the nature of plaintiff's injury.[18] Thus in many circumstances lay testimony that exposure to defendant's product caused plaintiff's illness or injury will not suffice to raise a fact issue of defendant's liability.[19]

1.4 RELATIONSHIP TO ENVIRONMENTAL LAW

Manufacturers of a wide array of potentially toxic products are subject to government licensing, regulation, enforcement, and penalties for derelictions. This licensing may be administered under the Environmental Protection Act (EPA), the Federal Hazardous Substances Act (FHSA), the Federal Insecticide, Fungicide and Rodenticide Act (FIFRA), or a variety of other laws and associated regulations. That body of law, i.e., public regulation of toxic harms to persons, property, or the environment, is often described as "environmental law".

A person's act, or failure to act, regarding a toxic substance may violate an environmental or occupational health statute or regulation. Most environmental claims involving toxic harm are brought by the government, be it the federal government, the state government, or a smaller public body such as a city, a county, or a town. These suits claim that defendant's conduct violated a statute or a regulation. A sprawling number of statutes may pertain to toxic wrongdoing. A partial list of such statutes, at the federal level, would include the FIFRA,[20] the Toxic Substances Control Act (TSCA),[21] the Surface Mining Control and Reclamation Act (SMCRA),[22] the National Environmental Policy Act (NEPA),[23] the Solid Waste Disposal Act (SWDA),[24] the Comprehensive Environmental Response, Compensation, and Liability Act (CERCLA),[25] the Occupational Safety and Health Act (OSHA),[26] the Federal Water Pollution Control Act (FWPCA),[27] and the Clean Air Act (CAA).[28] In addition, the numerous state environmental statutes provide mechanisms for imposing penalties or remediation requirements upon those responsible for environmental toxic harms.[29]

Unlike toxic tort suits, most environmental claims brought pursuant to statute by a public body are called "enforcement" or "penalty" actions. To remedy a toxic wrong, a state or federal agency might bring an action against the perpetrator for civil, or even criminal, money penalties. Significantly, and also unlike the operation of compensatory or punitive damages in a toxic tort claim, where the government prevails in a penalty or an enforcement action, any money penalties go to the government.[30]

To be contrasted, "toxic torts" describe individual, civil suits, seeking a remedy of money damages or injunctive relief, against an entity responsible for the harm. While official entities or agencies may be able to proceed against the person responsible for toxic harm, a toxic tort claim is the remedy available to you, to your neighbor, to a regional farmer, or to a local hospital patient, against the responsible party. While claims brought pursuant to statutes most often seek civil fines or criminal penalties, toxic tort claims are usually described as claims in negligence, implied warranty, strict tort liability, nuisance, or trespass and seek money damages commensurate with the harm defendant's conduct has caused. These remedies are often described as "common law" remedies, to distinguish them from statutory remedies that the state or federal government might have under federal statutes, or pursuant to state laws similar to these federal statutes.

An intentional workplace criminal act resulting in a toxic harm may create criminal liability. In addition to inviting criminal prosecution by the state, defendant's conduct may subject him to a civil toxic tort remedy.[31] In a criminal action against a person who knowingly pollutes the environment, a state or federal environmental authority might seek criminal sanctions, including fines or even jail. A toxic tort remedy, on the other hand, might be pursued by persons who suffered personal physical injury or property damage and who would sue the perpetrator for *civil* compensatory damages, injury, or property damage, resulting from the criminal conduct.

NOTES

1. As used throughout this work, "persons" means, without limitation, individual persons, business entities (including corporations), and public entities (including governmental subdivisions). Where person is used in its more individual sense, the text will use the terms "natural person[s]" or "individual persons[s]."
2. E.g., *Brower v. E.I. Dupont de Nemours & Co.*, Prod. Liab. Rep.(CCH) ¶ 12,640 (Idaho 1990) (damages resulting from farmer's use of herbicide precluded planting of crops other than wheat or barley for a time period substantially longer than contemplated at the time of purchase and initial application; held: claim barred by statute of limitations).
3. E.g., *Rushing v. Hooper-McDonald, Inc.*, 293 Ala. 56, 300 So. 2d 94 (Ala. 1974) (asphalt contamination).
4. The metes and bounds of punitive damages in toxic tort claims are discussed later in Chapter 18, and "citizen suit" claims under representative environmental statutes are described in Chapter 13.
5. E.g., *Sterling v. Velsicol Chemical Corp.*, 855 F.2d 1188 (6th Cir. 1988) (contamination of 242 acres used as a site for a landfill for by-products of manufacture of chlorinated hydrocarbon pesticides).
6. *Cf.*, *Shorter v. Champion Homebuilders Co.*, _____ F. Supp. _____, 1991 WL 212776 (N.D. Ohio 1991) (claim against homebuilders alleging injurious exposure to formaldehyde vapors, and alleging diverse personal injuries, including lethargy, headaches, and emotional harm).
7. E.g., *Bond v. Octagon Process, Inc.*, 745 F. Supp. 710, Prod. Liab. Rep. (CCH) ¶ 12,737 (M.D. Ga. 1990), *aff'd*, 926 F.2d 1573 (11th Cir. 1991), *cert. denied*, _____ U.S. _____, 111 S. Ct. 2855 (1991). In that suit, dismissed for want of *in personam* (personal) jurisdiction over defendant cleaning solvent manufacturer, plaintiff contracted Raynaud's phenomenon, described by the court as "the freezing of flesh in such a manner as to damage permanently blood vessels and cut off circulation . . . caused by a cold environment and the defatting and evaporative effects of chemical solvents, rather than any toxic effects of the solvent in question itself." Id., ¶ 12,737 at 38,280.
8. See Comment, "Not Just for Doctors: Applying the Informed Intermediary Doctrine to the Relationship Between Chemical Manufacturers, Industrial Employers, and Workers," 85 N.W.U.L. Rev. 562 (1991).
9. See generally *Sterling v. Velsicol Chemical Corp.*, 855 F.2d 1188 (6th Cir. 1988) (*reh. denied en banc*) (landfill disposal of hazardous liquid and dry chemical waste).
10. E.g., *Ellingwood v. Stevens*, 564 So. 2d 932 (Ala. 1990) (claim of inadequate shielding of patient's spinal cord during radiation therapy).

11. E.g., *Koslowski v. Sanchez*, 563 So. 2d 937 (La. App. 1990) (root canal filler use of N-2 paste containing 6.5% paraformaldehyde).
12. E.g., *Stead v. F.E. Myers, Co.*, _____ F. Supp. _____, (No. 89–169) (D. Vt. 1990) (well water contamination action involving polyaromatic hydrocarbons; suit brought in negligence, warranty, strict liability, and private cause of action under the Vermont Groundwater Protection Act).
13. 684 F. Supp. 852 (M.D. Pa. 1988).
14. E.g., *Graham v. Pittsburgh Corning Corp.*, 593 A.2d 567 (Del. Sup. 1990) ("Delaware law measures the duties owed in terms of reasonableness. One's duty is to act reasonably, as a reasonably prudent man (or entity) would. (citations omitted) One breaches that duty by not protecting against an event that a reasonably prudent man would protect against. Stated differently, one's duty encompasses protecting against reasonably foreseeable events.(citations omitted)." *Id.* at 568).
15. E.g., Dressen, et al., "A Study of Asbestosis in the Asbestos Textile Industry," Public Health Bull. No. 241 (1938). The Dressen study has been cited in leading decisions such as *Borel v. Fibreboard Paper Products Corp.*, 493 F.2d 1076, 1084 n.9 (5th Cir. 1973) (*reh. denied en banc*).
16. For example, defendant in a medical device suit where plaintiff alleges that use of defendant's intrauterine contraceptive device caused internal injuries might attempt to show that plaintiff's sexual activity, and not the intrauterine device (IUD), was the producing cause of the disease. Defendant in an asbestos case might produce evidence that plaintiff's habit of cigarette smoking, and not exposure to defendant's product, was the cause of plaintiff's lung disease.
17. The proposition is true whether a suit is to be heard by a jury, or by a judge as finder of fact.
18. Plaintiff's injury may include the *risk* of future illness. E.g., *Stead v. F.E. Myers, Co.*, _____ F. Supp. _____, (No. 89-169) (D. Vt. 1990) (expert testimony permitted on issues of (1) plaintiff's increased risk of contracting cancer as a result of exposure to polyaromatic hydrocarbon contamination of well water, and (2) the reasonable necessity of continued medical monitoring of plaintiffs for the purposes of early detection of future illness that might be caused by their exposure).
19. E.g., *Allen v. Roddis Lumber and Veneer Co.*, 796 S.W.2d 758 (Tex. App. 1990) (*error denied*) (lay [nonexpert] testimony that vapor emissions from furniture contained levels of formaldehyde that caused plaintiff's respiratory problems insufficient to raise factual issues; plaintiff's claim was one requiring expert testimony permitting inference of causation).
20. 7 U.S.C. § 136 *et seq.*
21. 15 U.S.C. § 2601 *et seq.*
22. 30 U.S.C. § 1201 *et seq.*
23. 42 U.S.C. § 4321 *et seq.*
24. 42 U.S.C. § 6901 *et seq.*
25. 42 U.S.C. § 9601 *et seq.*
26. 29 U.S.C. § 651 *et seq.*
27. 33 U.S.C. § 1251 *et seq.*
28. 42 U.S.C. § 7401 *et seq.*
29. E.g., Massachusetts Water Quality Standards, codified at 31 Code of Massachusetts Regulations 4.01 *et seq.* (1986).

30. Penalty moneys recovered may devolve to the government's general fund, or to remediation efforts authorized by statute.
31. For example, a criminal act resulting in death may result in a prosecution for homicide or manslaughter, to be brought by the state, while decedent's survivors may have a civil cause of action against the responsible party for compensatory damages for "wrongful death".

CHAPTER 2

Parties

CONTENTS

Parties

2.1 PLAINTIFFS

2.1.1 Generally

Potential claimants in a toxic tort may be individuals, legal representatives of those individuals, class groupings of individuals, business entities, or public representatives. Common to each claimant must be evidence that they purchased, used, consumed, or were foreseeably affected by the toxic substance.

The toxic tort claimant need not be an individual. Business entities, not limited to corporations, may prosecute toxic tort actions against manufacturers or suppliers whose product causes property damage, loss of property value, or other economic harm. Lastly, governmental entities, from states to municipalities or other political subdivision, may bring suits in trespass, public nuisance, or other common law claims against toxic polluters. In many of the above representative settings, a claimant's cause of action may be linked to a private right of action, called a "citizen suit", described in a related state or federal statute.[1]

2.1.2 Buyers, Users, or Consumers

The central goal of products liability remedies for toxic torts is the provision of compensatory damages to buyers, users, or consumers who suffer injury or loss caused by unreasonably dangerous products.[2] Suits on behalf of minors, or potential claimants who have died, may be brought by their legal representatives. Plaintiffs may bring actions on behalf of a class of similarly situated persons claiming injury or damage where the class representatives can demonstrate sufficient commonality of causation and damages.[3] The value of class action administration of mass toxic tort claims becomes all the more apparent where accumulated claims, if brought individually by injured persons, threaten to exhaust the financial resources of the responsible defendants, leaving numbers of injured persons with no realistic remedy.[4]

Individual plaintiffs range from workers exposed to a workplace toxin to homeowners exposed to urea-formaldehyde vapors emanating from building particleboard. Claims for personal physical injury or property damage arising from exposure to unreasonably dangerous products may be styled in negligence, strict products liability, warranty, or misrepresentation.

2.1.3 Citizen Suits Pursuant to Environmental Statutes

Under a variety of federal and state environmental statutes, individuals may sue defendants responsible for environmental harm for a wide range of non-compensatory relief. Called "citizen suits", these actions, brought under the auspices of a statute, authorize civil penalties to be paid to the state or federal government's general fund, certain response costs, injunctive relief, court costs, and fees. Citizen suit claims under the federal Clean Water Act, the Clean Air Act, and other illustrative statutes, are described in Chapter 13.

2.1.4 Employees and Other Handlers

A worker injured by exposure to a toxic product in the course of manufacture or handling may proceed against the manufacturer of that product upon sufficient proof that the named manufacturer supplied the product causing the injury or disease. For example, in *Celotex Corp. v. Tate,* a Texas appellate court held that plaintiff, the estate of a wallboard and plaster worker who died of pleural mesothelioma, produced sufficient evidence that Celotex Corp. manufactured the asbestos products that decedent regularly mixed with other substances to produce kalite, a soundproofing material.[5]

2.1.5 Bystanders

Persons other than those the manufacturer or seller intended would use, enjoy, or work directly with a toxic product are described as "bystanders". Bystanders may bring a toxic tort claim where they, or their property, suffer an injurious exposure to the defendant's product or process. A homeowner's personal, physical injury caused by the accidental environmental dispersal of chlorine gas from a damaged railway tank car may be redressed in a toxic tort claim against the manufacturer of the chlorine gas, the transporter, or even the railroad. As is described more fully later, success of the homeowner's claim may depend upon proof that one, or each, of these potential defendants was negligent, i.e., that they acted without due care under the circumstances. Where a defendant's acts result in a *trespass* upon plaintiff's land, plaintiff may recover damages irrespective of the care defendant exercised.

It is also possible that given the extraordinary risks involved in transporting toxic chemicals such as chlorine, that a claimant could proceed under the strict liability doctrine of abnormally dangerous activities. In such a claim, the plaintiff might prove liability against a transporter without having to prove negligence. Tort theories of liability, some of which require plaintiff to prove negligence and others that do not, are described more fully later.

The wide variety of tort suits involving asbestos-related injury provide additional illustrations of potential bystander liability. While persons who worked directly with asbestos products comprise the majority of asbestos-related personal injury claims, other plaintiffs have included shipyard, construction, or other co-workers whose exposure to respirable asbestos has been more remote. These bystanders have been permitted to recover damages upon a showing that their proximity to asbestos work, and the duration of their exposure, was a substantial contributing factor in the cause of their subsequent asbestos-related disease.[6]

While the duties of product sellers have been substantially enlarged to include a duty to bystanders, bystander plaintiffs face substantial difficulty in proving the suitability of punitive damages.[7]

2.1.6 Offspring

Certain examples drawn from pharmaceutical products liability cases illustrate that the spectrum of potential claimants is not confined to persons injured themselves by a product or a substance. For example, in DES litigation, a cause of action is recognized generally for children injured *in utero* by their mother's ingestion of DES during pregnancy. It is noteworthy that the class of potential claimants in those suits has not, to date, been extended to third generation claimants.[8] The DES cases suggest also that the cause of action brought by a person injured years before *in utero* does not even depend upon the mother's proof that she herself suffered compensable, physical injury. In the DES example, mothers who ingested the harmful drug certainly did suffer physiological change, with resulting gestational change to the fetus. In most cases, however, the mother did not suffer any diagnosable symptoms of disease.

2.1.7 Relational Plaintiffs

Where a person has suffered severe personal physical injury due to toxic exposure, in certain selective circumstances, close relatives have been permitted to bring tort claims in their own behalf. Some claims by relational plaintiffs claim injurious exposure to a toxin and consequent personal physical injury. Other claims do not posit personal physical injury, but rather emotional harm or distress caused by the injury or disease caused a close relative. As an example of the former claim, in some instances family members of asbestos workers have been permitted to recover where they have succeeded in demonstrating that asbestos fibers brought home on the person or clothing of their relative exposed them to levels of asbestos fibers that caused their disease.

Relational plaintiffs may also seek recovery for emotional harm or loss. In many jurisdictions, a close relative who sees a family member's injurious exposure to a toxic substance, and recognizes the potential for harm to another or to himself, may seek recovery for defendant's negligent infliction of emotional distress. Such claims do not rely upon the relative's proof that they suffered personal physical injury themselves, for in most instances they have not. Rather, the claim of the relational plaintiff is that their family member's injury or disease has caused them emotional harm or loss. The proof required to show compensable emotional harm varies dramatically from state to state.

In general terms, however, a relative not physically injured may have a tort suit in emotional distress, or in loss of consortium, or under both theories. Described more fully later,[9] a claim for loss of consortium states that the injurious incapacitation of one's spouse has deprived the other spouse of what the common law called ''consortium''. In its original and limited definition, consortium includes the support, satisfactions, and solace associated with the marital relationship. In most jurisdictions, therefore, only a husband or a wife may bring a consortium claim arising from incapacitation of their marital partner.

Significantly, some modern decisions have held that as marital sexual contact is only one component of the consortium claim, its absence in the relationships between and among other family members should not deprive them of consortium claims arising from their deprivation of the other-consortium-related qualities of family life. Courts so concluding have therefore permitted loss of consortium claims to be brought by children on the basis of incapacitation of a parent, or parents on the basis of the incapacitation of a child.

2.2 DEFENDANTS

2.2.1 Generally

A toxic tort claimant may seek money damages from any individual or other entity whose toxic product is a direct cause of their physical injury, property damage, or economic loss. Product manufacturers and other sellers are toxic tort defendants of first resort.

Depending upon the nature of their participation in the chain of distribution or marketing of a toxic substance, a variety of other entities may face potential liability. In some settings, liability may be imposed upon product designers, licensors,[10] trade associations, wholesalers or other distributors, and even health care providers, such as physicians or hospitals.[11] Persons responsible in toxic tort damages may also include those engaged in the storage or transportation of such products, or their application or dispersal. Toxic tort claimants have named environmental consultants in suits claiming that erroneous advice about toxic risks or appropriate disposition of toxins contributed to claimants' harm.[12]

Plaintiffs undergoing radiation therapy have pursued claims against health care providers and others for injuries claimed to have been caused by inadequate shielding of proximate body areas.[13] To be contrasted, in *Star v. Berridge,*[14] the New York Court of Appeals held that a landlord owed no duty of care to, and therefore would not be liable to, the occupant of an office next to another leased premises in which X-ray equipment was being operated.

Corporate officers may be held individually liable in tort for damages caused by toxic discharge "where the corporate officer participates in the tortious activity."[15] Factors considered in so concluding were summarized by one court as including "whether the person or corporation had the capacity to discover in a timely fashion the release or threat of release of hazardous substances; whether the person or corporation had the power to direct the mechanisms causing the release; and whether the person or corporation had the capacity to prevent and abate damages."[16]

A finding that a producer satisfies the definition of a successor for the purposes of potential liability may have implications for permissible *in personam* (personal) jurisdiction over the surviving entity. There is authority holding that a successor liable for the actions of the predecessor may be properly brought before any court in which personal jurisdiction would have been appropriate as to the predecessor.[17]

While plaintiffs seeking toxic tort money damages must ordinarily identify the specific defendant or defendants responsible for the harm, in many cases the passage of time, absence of records, or complex questions of medical or scientific causation make identification of particular defendants impossible. In a growing number of such cases, courts have crafted a remedy permitting claimants to name as defendants a group of persons responsible for marketing a product. Depending upon the state substantive law applied to a particular suit, such group defendants may then exculpate themselves by proving that they did not market the product responsible for plaintiff's loss, did not sell products that would be expected to find use in plaintiff's locale, or did not, by product design or marketing, otherwise contribute to the risk to which plaintiffs were exposed.[18]

NOTES

1. In most settings, toxic tort claimant need not rely upon a so-called statutory right to sue. The evidentiary effect of defendant's compliance with, or violation of, pertinent environmental, workplace, or product safety statutes is discussed in Chapter 14, *Defenses*.
2. Toxic tort products liability remedies in negligence and in strict products liability are discussed in Chapters 3, 5 and 7.
3. E.g., *Sterling v. Velsicol Chemical Corp.*, 855 F.2d 1188 (6th Cir. 1988). In the appeal of this suit brought by a class of persons who lived in the vicinity

of a landfill, and who alleged that defendant's waste disposal in that landfill contaminated well water and groundwater, the appeals court approved class certification under Fed. R. Civ. Proc. 23(b)(3), commenting: "[A] class action in the instant case avoided duplication of judicial effort and prevented separate actions from reaching inconsistent results with similar, if not identical, facts." *Id.*, 855 F.2d at 1197.

4. E.g., *Loper v. Eagle-Picher Industries, Inc.*, 1990 Toxics Law Rptr. (BNA) 494 (Sept. 12, 1990) (No. CV 87-1383 (E.D.N.Y.) (motion filed 7/23/90) (defendant manufacturer of asbestos products seeking class action certification, claiming that "claims are made by numerous persons against a fund insufficient to satisfy all claims," citing Advisory Committee Notes to Fed. R. Civ. Proc. 23.
5. *Celotex Corp. v. Tate,* 797 S.W.2d 197, Prod. Liab. Rep. (CCH) ¶ 12,673 (Tex. App. 1990).
6. See generally Comment, "Not Just For Doctors: Applying the Learned Intermediary Doctrine to the Relationship Between Chemical Manufacturers, Industrial Employers, and Employees," 85 N.W.U.L. Rev. 562 (1991).
7. E.g., *Eagle-Picher Indus., Inc. v. Balbos,* 84 Md. App. 10, 578 A.2d 228 (Md. App. 1990) (failure of two shipyard workers who died of mesothelioma to prove manufacturer's wanton or reckless conduct where plaintiffs were workplace bystanders, and not workers actually using the products).
8. *Enright v. Eli Lilly & Co.*, 77 N.Y.2d 377, 586 N.Y.S.2d 550, 570 N.E.2d 198, Prod. Liab. Rep. (CCH) ¶ 12,743 (1991), *cert. denied* _____ U.S. ___ (1991).
9. See Chapter 18, *Damages.*
10. Cf., *Peterson v. Sealed Air. Corp.*, 902 F.2d 1232, 16 Fed. R. Serv. 3d 920 (7th Cir. 1990), where the federal appeals court held that irrespective of whether a trademark licensor may ultimately be made a party defendant in a toxic tort claim, that service upon the registered agent of a manufacturer of methyl isocyanate naming the manufacturer's subsidiary, formed to hold and license the trademark, sufficed to put the manufacturer on notice of the claim.
11. E.g., *Blake v. Mead Johnson & Co.*, 1990 Toxics Law Rptr. (BNA) 824 (November 28, 1990) (Ill. Cir. Ct. No. 86L26184) (two hospitals held jointly liable along with manufacturer of asthma medication for physicians' failure to monitor heightened levels of theophylline in infant's blood).
12. *Stoppleworth v. Refuse Hideaways, Inc.*, 1990 Toxics Law Rptr. (BNA) 817 (Nov. 28, 1990) (Wis. Cir. Ct. No. 90-2245).
13. *Ellingwood v. Stevens,* 564 So. 2d 932 (Ala. 1990) (allegation that spinal cord was inadequately shielded from radiation therapy).
14. 568 N.Y.S.2d 904, 571 N.E.2d 74 (N.Y. 1991).
15. *United States v. Carolina Transformer Co., Inc.*, 739 F. Supp. 1030, 31 E.R.C. (BNA) 2120, 2126 (E.D.N.C. 1989) (PCB contamination, citing *Polo Fashions v. Craftex, Inc.*, 816 F.2d 145, 149 (4th Cir. 1987)).

16. *United States v. Carolina Transformer Co., Inc., id.*, 31 E.R.C. (BNA) 2120 at 2126. Finding individual common law liability, the court stated: "The facts clearly establish that Dewey and Kenneth Strother knew about the hazardous waste problem and willfully failed to correct it. Accordingly, both . . . are liable under the common law theory of individual liability."
17. *Simmers v. American Cyanamid Corp.*, 576 A.2d 376 (Pa. Super. 1990) (predecessor's production of Thorotrast, a radioactive contrast dye used in diagnostic X-rays).
18. Joint and market share liability theories and application are discussed in Chapter 10.

CHAPTER 3

Overview of Toxic Torts — The Theories Compared

CONTENTS

Overview of Toxic Torts — The Theories Compared

3.1 GENERALLY

The toxic tort claimant will lodge claims against one or more defendants on multiple causes of action. For example, a homeowner claiming that water runoff from a nearby asphalt-producing facility has polluted her pond might bring claims in (1) nuisance, alleging that the contamination of the pond interferes with her quiet enjoyment of her home and property; (2) trespass, alleging that the asphalt effluent constituted an actionable invasion of her possessory interest in her property; and (3) negligence, alleging that the asphalt producer failed to exercise due care in conducting its operations so as to prevent, to the extent practicable, the claimed intrusion and contamination.[1]

3.1.1 Nuisance and Trespass

A defendant's actions resulting in contamination or pollution of another's property, or the injurious exposure of another to a toxic chemical or substance, may give rise to a claim in *private* nuisance. Where defendant's conduct interferes with or harms a right common to the larger public, a public authority may bring suit for *public* nuisance. In certain limited circumstances where plaintiff's harm, although not to his *own* property, is *individual* and distinct from the harm suffered by the broader public, an individual may bring suit in his own behalf in public nuisance.

The *intent* of defendant's conduct that plaintiff must show varies from nuisance to trespass. In trespass, plaintiff does not need to show that defendant *intended* that his product or process invade or contaminate plaintiff's land. Regardless of defendant's good faith, or caution, or innocent mistake, a claim for at least nominal damages will be available to a plaintiff suffering toxic contamination of property. This legal solicitude toward the interests protected by the trespass cause of action is due to "the high social value traditionally placed upon the inviolability of a person's real property, and particularly of his home."[2]

In nuisance, on the other hand, plaintiff must show that defendant's interference with the use or enjoyment of property was either "intentional and unreasonable", or, if unintentional, was negligent, reckless, or could be classified as an "abnormally dangerous" activity.[3]

While a claim in trespass usually requires that plaintiff demonstrate that defendant's pollutant or contaminant has settled on or infiltrated plaintiff's property, in private nuisance, plaintiff does not have to prove that there was a physical *invasion* of the property. Rather, plaintiff needs only demonstrate that defendant's conduct unreasonably interfered with plaintiff's "private use and enjoyment of [his] land."[4] The *invasory* (physical interference) requirement of trespass, and absence of that requirement for a private nuisance claim, has greater significance outside of the toxic tort area, for most toxic tort claims arising from contamination or pollution involve a determinable *invasion* of property, even if the substance itself is invisible. For example, contamination by PCBs or by radiation is not visible to the naked eye, but can be considered, nonetheless, a physical invasion of plaintiff's property. Plaintiff may also frame a claim in *continuing* trespass,[5] which, upon sufficient evidence, "confers on the possessor of the land an option to maintain a succession of actions based on a theory of continuing trespass, or to treat the continuation of the thing on the land as an aggravation of the original trespass."[6]

The consequences of this are that in the realm of toxic torts, plaintiff will often have a claim in both private nuisance *and* in trespass. The distinctions between the two claims become significant in the description and measure of damages plaintiff seeks under each claim, respectively. For private nuisance, plaintiff asks for damages that will compensate for the interference with the use and enjoyment of the land. For example, a property owner's apple orchard mistakenly contaminated by aerial pesticide spraying over nearby land might seek nuisance damages for both (1) any period of time the owner had to reside elsewhere in order to permit the contaminant to dissipate and (2) any economic harm suffered by damage to the apples or necessary delay in harvesting. The orchard owner's damages in trespass, on the other hand, do not require that he show actual loss or damage. If the supposed aerial spraying did not require a temporary residence elsewhere, and if it neither harmed nor delayed the harvesting or sale of apples, the owner would still be entitled to *nominal* damages because defendant's spraying operation *technically*, although innocuously, invaded the property.

3.1.2 Negligence

In negligence, plaintiff must demonstrate that defendant acted without due care under the circumstances and that this conduct was a direct cause of plaintiff's injury or loss. Decisions applying the common law negligence standard require that plaintiff, in her prima facie case, produce evidence that defendant (1) owed plaintiff a duty of due care; (2) breached that duty; (3) the breach of duty was the proximate or legal cause of plaintiff's injury or loss; and (4) that plaintiff suffered compensable damages.

The lack of an existing testing mechanism or precautionary device, available at the time the product is sold, that would render product exposure duly safe will not preclude negligence liability if the provider did not utilize tests or means of distribution that would have reduced, if not eliminated, the risk.[7]

Generally speaking, a defendant owes a duty to those whose physical safety or property would be put at an unreasonable risk if defendant failed to act with due care. The test is one of foreseeability, i.e., would a reasonable person in defendant's position foresee a measurable risk to plaintiff. Plaintiff who successfully establishes that defendant owed a duty of due care must then show that defendant breached that duty. In deciding if defendant has breached his duty, most courts apply one or another variants of a negligence risk utility analysis. A conventional expression of such risk-utility weighing is this: where the burden upon defendant of taking precautionary measures that would have prevented the harm is less than the *multiple* of (1) the probability that the harm will occur absent additional precautions and (2) the magnitude of the harm should it occur, defendant has breached a duty to plaintiff.[8]

In negligence actions against manufacturers of toxic substances, many courts employ a balancing test for determining whether defendant breached its obligation of reasonable care.[9] The level of care required of a seller, supplier, or contractor handling toxic substances is greater where (1) the likelihood of harm is great or (2) even if the likelihood of harm is small, the gravity of the harm, should it occur, is great. Under a negligence standard, the care required of a defendant handling a product that has not previously been known to cause toxic harm, or is known to cause toxic harm only in an extremely small segment of the population, is less than the degree of care required of one handling a product known to cause harm invariably to any person or property exposed to it. Likewise, the degree of care of a defendant handling a product that, if it injures anyone at all, creates a minor temporary skin rash is lower than that which the law requires of the defendant handling a toxic product that, upon accidental exposure, causes serious injury or illness.

Significantly, evaluation of the duty of care, central to evaluating a negligence claim, is not a factor in claims brought in nuisance, trespass, warranty, strict products liability, or liability for abnormally dangerous activities. In claims brought under those theories, (1) the risks created by a toxic product, (2) the failure of the product to perform as represented or expected, or (3) the product's intrusion upon or damage to plaintiff's property can, where injury or loss occurs, mean liability for the defendant even where the defendant has exercised due care to prevent the harm.

3.2 WARRANTY

There are three types of warranty claims available to the toxic tort claimant. Depending upon the circumstances of the harm, plaintiff may claim that defendant breached: (1) the product's express warranty; (2) the implied warranty of merchantability; or (3) the implied warranty of fitness for a particular purpose. In warranty claims, plaintiff's injury or loss may be severe, involving personal injury or property damage, or it may be only the lost "benefit of the bargain", in that the product failed to perform as warranted.

Express warranty claims, and those of breach of the implied warranty of merchantability, may only be brought against a defendant who sells those products in the ordinary course of their business. Claims for breach of the implied warranty of fitness for a particular purpose can be brought against one who is not a merchant of such goods, although as a matter of experience, practically all warranty claims of any description are brought against manufacturers or merchants who are engaged regularly in the sale of the subject goods.

To prove liability for breach of express warranty,[10] a plaintiff must show that (1) defendant gave an explicit assurance about the safety or performance of the product; (2) the assurance or representation proved incorrect; and (3) that plaintiff suffered injury or loss as a consequence. A representation that gives rise to an express warranty may take the form of an advertisement, a label, instructions for use, or even spoken assurances of performance or safety that defendant's sales personnel or representatives make to a buyer.

Even where the manufacturer or seller is silent as to the safety or qualities of a product, any seller *impliedly* assures the merchantability and reasonable safety of any product sold. Unless this implied warranty of merchantability is disclaimed conspicuously, a seller impliedly warrants that a product is "fit for the ordinary purpose for which such goods are used."[11] Significant for the purposes of sales of toxic products, the language "fit for the ordinary purpose" has been interpreted to mean that the product can be used in a reasonably safe manner. A product that creates an unreasonable risk of personal injury or property damage when used in its intended or reasonably foreseeable way is *not* "fit for its ordinary purpose", i.e., it is unmerchantable. Although a product performs its intended function adequately, it may be considered nonmerchantable if it poses a risk of personal injury or property damage greater than would be expected by the ordinary consumer.

For example, while asbestos products may be excellent, even unequaled, in their application as insulating material or fireproofing, such products can be considered unreasonably dangerous, and nonmerchantable, where their fabrication or installation permits release of respirable fibers. Similarly, a cleaning solvent may perform well in its designated task as a tar remover, but be nonmerchantable if not accompanied by warnings and instructions adequate to alert the user of risks of inhalation, and advising of ventilation or other means of reducing such a hazard.

For the distinctions with claims in negligence, the implied warranty claim enjoys similarities as well. A plaintiff who proves defendant's negligence liability for a toxic harm should almost always prevail in the count against defendant for breach of the implied warranty of merchantability. Consistent therewith, it has been held that a jury verdict finding a manufacturer of asbestos products liable in negligence, but not in breach of the implied warranty of merchantability, requires reversal.[12]

Although warranty claims are technically grounded in the law of contracts, they enjoy so substantial a relation to tort claims that in most jurisdictions a plaintiff's tort claim for injury caused by a defendant's product will also give rise to one or more warranty claims. Special warranty rules do, however, apply to issues such as (1) plaintiff's ability to recover damages for economic loss alone, in settings where plaintiff has suffered no personal physical injury or damage to property other than the warranted product itself; (2) the ability of persons other than the person who purchased the product, or the vendee's family members, to recover damages; and (3) a seller's ability to disclaim warranty liability, or to limit warranty damages. These issues are discussed in Chapter 6.

A seller may also breach the implied warranty of fitness for a particular purpose where a purchaser, seeking a product for a specialized purpose, relies upon the seller's knowledge or expertise in the selection, and the product's subsequent failure to perform as expected results in harm or loss to the buyer.[13] The buyer's actual reliance upon the seller's expertise must be shown; therefore, the purchaser who is as knowledgeable as the seller of a product's performance or risks cannot rely upon this implied warranty.

In addition, the product must be one that is to be used for a particular purpose, i.e., a purpose out of the ordinary. Thus a homeowner asking a knowledgeable seller for a recommendation of an enamel paint that might be used safely in painting a child's playroom might recover in the breach of the implied warranty of fitness for a particular purpose if chips of the paint suggested, and used, were highly toxic, with consequent injury to a child in the buyer's home. Another home painting example of the creation of the implied warranty of fitness for a particular purpose would be the homeowner seeking indoor paint for a basement recreation room subject to chronic dampness. Where the buyer communicates to the seller the need for a paint that is particularly suited to withstand sustained dampness, the sale of a paint demonstrably unfit for this application would breach this implied fitness warranty.

Claims for breach of express warranty or breach of the implied warranty of merchantability are available only against one who sells such products in the ordinary course of their business. To be contrasted, a claim for breach of the implied warranty of fitness for a particular purpose may be brought against any seller.

3.3 STRICT TORT LIABILITY

It is accepted generally that in strict tort liability, the toxic tort plaintiff must prove three elements: "(1) a defect existed in the product (2) which created unreasonable danger for the plaintiff and (3) which caused injury to the plaintiff."[14] Unlike a claim brought under a negligence theory, the focus of the strict tort liability remedy is the relative safety of the product, not the caution exercised by the defendant. Where the product sold is in an unreasonably dangerous and defective condition, and where that condition is the proximate cause of plaintiff's injury or loss, defendant may be liable in strict tort.

In warranty claims, the class of permissible claimants may be limited by application of state by state variations of UCC 2-318, by conspicuous disclaimers, or by limitations of remedy. A suit in strict tort liability, on the other hand, may be brought by any person or entity the seller might reasonably foresee would use, consume, or be affected by the product if sold in an unreasonably dangerous condition. Furthermore, courts are more reluctant to impose disclaimers or remedy limitations with regard to tort claimants than with regard to plaintiffs pursuing claims in breach of warranty.

The authors of Restatement, Second, Torts § 402A, the benchmark standard for strict liability claims, reserved judgement on whether these claims might be brought by bystanders, i.e., persons who did not themselves wittingly use or consume the product. Notwithstanding the absence of an explicit inclusion of a bystander remedy in § 402A, however, in practically every state the decisions under this section have held that bystanders capable of demonstrating their injurious exposure to the product in its ordinary use may recover in strict tort liability.

In claims in negligence, nuisance, trespass, and liability for abnormally dangerous activities,[15] a substance giving rise to toxic harm does not need to have been reduced to consumable form in order for the resultant injury to create a cause of action. For example, the escape of dangerous levels of radiation from a radiological diagnostic center, and the consequent contamination of a nearby public area, might give rise to claims in negligence or public nuisance even though the radiation could not, in that form, be considered a consumer product. To be contrasted, in strict products liability, as in warranty, the product or substance must be in useable or consumable form, or at least approaching that condition. A product not technically in consumable form, but the emission of which might give rise to a claim in strict products liability, would be bulk fuel, such as propane, prior to its infusion into storage at an individual residence. As in the principal warranty claims (excluding the implied warranty of fitness for a particular purpose), strict products liability for sale of a defective product may only attach to one who sells such products in the ordinary course of their business.

While technically it is the condition of the product, and not the prudence or negligence of the actor, that underlies the strict liability claim, the discussion to follow in Chapter 7 will describe how in the evaluation of whether a product is sold in an unreasonably dangerous condition, the court or the jury is invited to consider various factors that seem indistinguishable from those of concern in negligence evaluations. Such factors may include, among others, whether the product's risks were known or knowable and whether the manufacturer could have designed or formulated the product in a safer condition without materially lessening its usefulness or availability. The key to strict liability analysis is that the manufacturer of a product is presumed to be an expert in matters affecting design, formulation, and marketing matters and the knowledge of an expert in such fields will be *imputed* to the manufacturer irrespective of whether the individual defendant was aware of the hazard.

3.4 INTENTIONAL TORTS

3.4.1 Effect of Workers Compensation Bar

A worker injured by workplace exposure to a toxic substance might, depending upon the circumstances of the injury, consider lodging tort claims against the manufacturer or seller of the substance, as well as against his or her employer, if the employer's substandard conduct contributed to the risk of injury. Worker's compensation statutes generally bar consideration of the latter claim, confining employee claims against employers to the categories and amounts of pecuniary loss the injured employee can prove. The workers compensation bar of employee-employer suits is intended to preclude claims based upon the lack of caution or negligence of an employer, and thus the bar to employer-employee actions will be preserved where the injured employees are unable to demonstrate that the employer engaged in willful disregard of an unreasonable risk to the health of employees.

To avail himself of a tort action against the employer, the employee must show that the employer, knowing of a substantially certain harm to the employee, intentionally required the employees continued exposure to it.[16] Illustrative of the exception to the worker's compensation limitations is one suit in which an Ohio court held that an employee raised a triable issue of his employer's intentional and wrongful conduct with evidence that, although employees received certain warnings about the risks of dermal and respiratory exposure to a workplace chemical, the employer had additional knowledge of the risks and denied the existence of such risks in response to employee inquiries.[17] Conversely, municipal workers have been held to be precluded from a tort action against their employer following their exposure to asbestos in the course of a removal project where the employees were unable to show that the municipality knew that the material they asked be removed contained asbestos until the workers were already exposed to the hazard.[18]

3.4.2 Fraud or Deceit

In many jurisdictions a seller who publishes misrepresentations of material fact about a product will be liable without fault, i.e., without the need for plaintiff to prove that the misrepresentation was negligent or intentional.[19] In toxic tort claims as elsewhere, plaintiff claiming fraud must plead the fraud with particularity. The practice rules in many jurisdictions require that "the circumstances constituting fraud . . . be specifically averred."[20]

Illustrative of this rule is the suit in which a producer of the phosphate slag component of building blocks used to construct homes was found not liable in fraud when a homeowner later discovered radon gas emissions where the court found that the producer neither knew nor had reason to know of the hazard at the time the blocks were sold.[21] Similarly, fraud claims against manufacturers of tobacco products that omit to specify "the time, the place, the substance of the false representations, the facts misrepresented, and the identification of what was procured by fraud" have been dismissed for failure to state a claim.[22]

NOTES

1. Toxic tort claims available in other situations might, on these facts, be unavailable to our hypothetical homeowner. A warranty claim might fail for want of a sale; a breach of the implied warranty of habitability might fail for the additional reason that the asphalt producer, unlike the homeowner's direct vendor or building contractor, has warranted nothing concerning the home; a strict tort liability products action might fail because, inter alia, at the time of the damage, the product was not reduced to consumable form, i.e., was not at that time a "product"; and an action claiming liability for an abnormally dangerous activity might fail should it be shown that asphalt production can, with the exercise of reasonable care, avoid the claimed contamination risks and damage. Warranty, strict tort, and abnormally dangerous activity liability claims are discussed in Chapters 6, 7, and 8, respectively.
2. Christie and Meeks, *Cases and Materials on the Law of Torts* 52 (2d ed.) (West 1990).
3. "One is subject to liability for private nuisance" if the invasion of another's interest in the use and enjoyment of their property is either "(a) intentional and unreasonable, or (b) unintentional and otherwise actionable under the rules controlling liability for negligent or reckless conduct, or for abnormally dangerous conditions or activities." Restatement, Second, Torts § 822.

   The law pertaining to proof of negligence, and the criteria for deciding where conduct constitutes an "abnormally dangerous activity", is summarized below in this chapter, and is also discussed in Chapters 5 and 8.
4. Restatement, Second, Torts § 822. E.g., *Cooper v. Armstrong Rubber Co.*, 29 E.R.C. (BNA) 1102, 1111 (S.D. Miss. 1989) (finding plaintiff's claim of nuisance arising from defendant's maintenance of waste sites "at least as detailed" as those countenanced in *Vicksburg Chemical Co. v. Thorness,* 355 So. 2d 299 (Miss. 1978)).

5. "The actor's failure to remove from land in the possession of another a thing he has tortiously . . . placed on the land constitutes a continuing trespass for the entire time which the thing is on the land[.]" Restatement, Second, Torts § 899, comment d.
6. Id., Restatement, Second, Torts § 899, comment d.
7. E.g., *Doe v. Univ. Hosp. of N.Y. Univ. Med. Center,* 148 Misc. 2d 756, 561 N.Y.S.2d 326 (N.Y. Sup. 1990) (hospital could be liable in negligence for patient's infection with AIDS virus from blood transfusion incident to surgery even where transfusion given prior to dispositive screening test for AIDS virus if hospital did not utilize then-extant screening measures and safeguards).
8. Additional discussion of plaintiff's prima facie case in negligence appears in Chapter 5.
9. See, e.g., *Graham v. Pittsburgh Corning Corp.*, 593 A.2d 567 (Del. Sup. 1990), a failure to warn action brought against the manufacturer of asbestos products. There the trial court states: "Some writers and judges have applied a test which balances the severity of harm which might result from use of the product against the burden of protective action by the manufacturer. *Westinghouse Electric Corp. v. Nutt,* [407 A.2d 606 (D. C. App. 1979)], citing 5 Harper & James, The Law of Torts 28.4 (1956). See also Noel, Manufacturer's Negligence of Design or Directions for Use of Product, 71 Yale L.J. 816 (1962); [1] Madden, [Products Liability (2d ed. 1988)] § 4.2." *Id.* at 569.
10. U.C.C. § 2–313.
11. U.C.C. § 2–314 (2) (c).
12. *Caccavale v. Raymark Industries., Inc.*, 404 Mass. 93, Prod. Liab. Rep.(CCH) ¶ 12,108 (Mass. 1989) ("The jury's answers to special questions stating that Raymark was negligent, but not in breach of warranty, were inconsistent as a matter of law." *Id.* at 35,247.). The court noted that an exception to this general proposition might be had in facts supporting liability for negligent failure to give a post-sale warning. *Id.* at 35,247–35,248 (citing *Hayes v. Ariens Co.*, 391 Mass. 407, 410, Prod. Liab. Rep. (CCH) ¶ 10,030 (1984).
13. U.C.C. § 2-315.
14. *Dillon v. Fibreboard Corp.*, 919 F.2d 1488, Prod. Liab. Rep. (CCH) ¶ 12,706 at 38,153 (10th Cir. 1990).
15. Liability for abnormally dangerous activities, and its underpinning in Restatement, Second, Torts §§ 519–520, is discussed in Chapter 8.
16. See *Van Fossen v. Babcock & Wilcox Co.*, 36 Ohio St. 3d 100, 522 N.E.2d 489 (1988), where the court states that for an employee to establish "intent" he must establish: " . . . (1) [K]nowledge by the employer of the existence of a dangerous process, procedure, instrumentality or condition within its business operation; (2) knowledge by the employer that if the employee is subjected to such . . . then harm to the employee will be a substantial certainty and not just a high risk; and (3) that the employer, under such circumstances, and with such knowledge, did act to require the employee to continue to perform the dangerous task." *Id.*).

17. *Richards v. C. Schmidt Company,* 54 Ohio App. 3d 123, 1989 WL 225788 (Ohio App. 1989). Reversing summary judgment on the employee's intentional tort claim against his employer, the court explained: ''According to the plaintiffs, Schmidt failed to pass along the information in the chemical safety data sheets to its employees. Dennis testified that Schmidt encouraged him to wash himself with the chemicals provided by Ashland. He further testified that Schmidt failed to provide protective clothing and equipment[.] . . . Schmidt also repeatedly assured Dennis that the chemicals would not harm him.''
18. *DeMoss v. City of Coer D'Alene,* 795 P.2d 875 (Idaho 1990).
19. Restatement, Second, Torts § 402B, ''Misrepresentation by Seller of Chattels to Consumer'', provides: ''One engaged in the business of selling chattels who, by advertising, labels, or otherwise, makes to the public a misrepresentation of a material fact concerning the character or quality of a chattel sold by him is subject to liability for physical harm to a consumer of the chattel caused by justifiable reliance upon the the misrepresentation, even though (a) it is not made fraudulently or negligently, and (b) the consumer has not bought the chattel from or entered into any contractual relation with the seller.
20. E.g., Ind. Trial Rule 9 (B).
21. *Wayne v. Tennessee Valley Authority,* 730 F.2d 392, 396 (5th Cir. 1984) (''Even considering the affidavit [submitted by TVA] in the light most favorable to the Waynes, it indicates that TVA did not know or have reason to know of the danger of radon gas in 1969, when the blocks at issue were sold, and did not know or have reason to know of such danger until near the end of the 1970's, when TVA stopped the sale of phosphate slag and told the public why it was doing so.'').
22. *Rogers v. R. J. Reynolds Tobacco Co.,* _____ Ind. App. _____, Prod. Liab. Rep. (CCH) ¶ 12,572 at 37,533 (Ind. App. 1990).

CHAPTER 4

Nuisance and Trespass

CONTENTS

Nuisance and Trespass

4.1 NUISANCE, GENERALLY

Defendant's conduct may create an actionable nuisance where it either interferes with (1) a public right or (2) a private party's interest in the use or enjoyment of land. The complainant in private nuisance seeks to protect his own, current interest in the quiet enjoyment of property. The plaintiff need not own the property; he need only be a lawful occupant. For a suit in public nuisance, on the other hand, the complainant need not have a property interest in any property affected by defendant's conduct.

4.1.1 Negligence and Trespass Distinguished

Unlike a claim in negligence, which requires proof that defendant's conduct was somehow substandard, the plaintiff alleging a private nuisance need not show that the actor creating a nuisance acted without due care under the circumstances. For example, in *Morgan v. High Penn Oil Co.*,[1] an adjoining landowner claimed that defendant's oil refining operation "emitted nauseating gases and odors in great quantities, which invaded plaintiff's land and other tracts of land within a distance of two miles, in such density as to render persons of ordinary sensitiveness uncomfortable and sick." Liability in nuisance could be found, the North Carolina Supreme Court stated, against one "who intentionally creates or maintains a private nuisance . . . regardless of the degree of care or skill exercised by him to avoid such injury."[2]

A claim in trespass ordinarily seeks damages for an intrusion onto a property, i.e., the pollution or contamination of plaintiff's property by particulate or subparticulate matter. In contrast, claims that defendant's conduct creates conditions of noise, lights, odor, or vibration that interfere with plaintiff's quiet enjoyment of the property ordinarily sound in nuisance.[3] Hence, where plaintiff's claim involves a physical intrusion upon his property, and where that intrusion is more than a transitory one, plaintiff may plead trespass.

Consistent with this general rule, one court distinguished nuisance from trespass on the basis of the duration of the interference. A fleeting or temporary interference with plaintiff's right to enjoy his property might be categorized as nuisance, while invasion of the property by the same substance or matter, if chronic, or where the contamination was more difficult to eliminate, could be trespass. On this rationale, the court stated that "airborne particles [that are] transitory and quickly dissipate, [and therefore] do not interfere with a property owner's possessory rights [are] properly denominated nuisances."[4]

4.2 ELEMENTS

Private nuisance may be described as an "unreasonable interference with the use or enjoyment of a property interest."[5] A toxic intrusion upon the land of a property owner will, therefore, constitute nuisance if it unreasonably interferes with the occupant's capacity to use, or quietly enjoy, the property.

The distinction between "use" and "enjoyment" is not semantic, as some low-level airborne pollution may not altogether prevent an occupant from using his property, but may, nevertheless, interfere with his enjoyment of it, as might a variety of intrusions by odor.[6] At common law, it is generally accepted that "smoke, offensive odors, noise or vibrations" that "materially interfere" with the possessor's "ordinary comfort" may constitute a nuisance.[7]

A toxic nuisance might take a wide range of forms. The pollution of a residence's well water would interfere both with a resident's use *and* enjoyment of a property. A sulfurous smell emanating from a business in a city's downtown financial district might constitute a nuisance if emitted every morning for several hours, but might not if it were discernable each day only in the 2 hours before dawn, when the affected office buildings and sidewalks alike were largely empty.

Where the harm to plaintiff's land is substantial, many courts have declined to permit defendant to exculpate itself by showing that the value of its conduct outweighed the gravity of any harm to plaintiff.[8] Some courts have adopted a "comparative injury" analysis.[9]

4.3 NATURE OF THE INTEREST INTERFERED WITH

The particular use to which a property is put, and the sensitivities of the persons using the property, may be a factor in evaluating if defendant's conduct constitutes an *unreasonable* interference that rises to the level of a nuisance. For example, if the morning sulfurous smell, described above, enveloped the premises of a nursery school playground, or a retirement residence, the proprietors of either could argue plausibly that the odor interfered substantially with their use of the properties for those purposes. If, on the other hand, the odor affected only adjoining properties engaged in smelting operations, any interference with either use or enjoyment might be too insubstantial to warrant a remedy in private nuisance.

Some decisions distinguish nuisance per se from nuisance per accidens. A nuisance per se would be any act that constitutes a nuisance "at all times and under any circumstances,"[10] such as, e.g., the permanent or chronic contamination of plaintiff's property. Nuisance per accidens, on the other hand, requires the fact finder's evaluation of whether, under all the surrounding circumstances . . . [defendant's action] substantially interferes with [plaintiff's] comfortable enjoyment[.]"[11] Successful nuisance claims per accidens have gained orders in abatement or damages for invasions by particulate matter such as limestone dust.[12]

An owner or occupier's departure from a premises will not preclude a subsequent action in nuisance where the interference with another's rights of enjoyment of the property is of a continuing nature. A landowner who has polluted or contaminated a property and then sold or leased it to another may remain liable "for the continuation of the nuisance" after the transfer, until the transferee "discovers the condition and has reasonable opportunity to abate it."[13] Although the vendor may be liable in nuisance to holders of rights to other property, the vendee cannot himself sue the vendor in nuisance. The federal appeals court in *Philadelphia Electric Company v. Hercules, Inc.*[14] so concluded, reasoning that a landholder's duty not to unreasonably interfere with use and enjoyment of land is owed to the landholder's *neighbors* and not the transferee of the land.[15] Thus, unless the vendee can show that the seller fraudulently concealed a contamination nuisance, the buyer of a property may be assumed to have realized that a property's purchase price "reflect[ed] the possibility of environmental risks, even if the exact condition . . . was not discovered."[16]

4.4 PUBLIC NUISANCE

Public nuisance is defined widely as "an unreasonable interference with a right common to the general public."[17] The Restatement, Second, Torts § 821B states that "circumstances" that might give rise to the conclusion that defendant's activity creates an *unreasonable* interference with a public right include evaluation of: "(a) whether the conduct involves a substantial interference with the public health, the public safety, the public peace, the public comfort or the public convenience, or (b) whether the conduct is of a continuing nature or has produced a permanent or long-lasting effect and, to the actor's knowledge, has a substantial detrimental effect upon the public right." It can be seen readily that the protections within the public nuisance remedy may pertain to toxic harms where a wide and accidental dispersal of toxins affect the public health, comfort, or convenience.

Defendant's conduct may create a cause of action in public nuisance even where "neither the plaintiff nor the defendant acts in the exercise of private property rights."[18] For example, in *Burgess v. M/V Tomano,*[19] commercial clam diggers and fishermen were permitted to pursue a public nuisance claim, premised on an off coast oil spill, even though it was "uncontroverted" that "the right to fish or to harvest clams in Maine's coastal waters is not the private right of any individual, but is a public right held by the State 'in trust for the common benefit of the people'."[20]

For the most part, the public nuisance remedy is enforced by a government body, such as a town, on behalf of the public. For example, depending upon the common law of a particular jurisdiction, a municipality might lodge a claim in public nuisance against a manufacturing facility discharging chemical effluent from the plant contaminating a nearby lake, killing aquatic life, and precluding recreational sports and swimming. If the contamination was substantial, and the effect would take measurable time and expense to eliminate, the claim in public nuisance would state that the manufacturer's conduct created an unreasonable interference with rights common to the public. The rights interfered with would be the public health, public safety, and the public convenience associated with fishing and aquatic sports.

While suits in public nuisance are usually brought by public bodies, under certain circumstances a private individual may sue in public nuisance. An individual may sue another in public nuisance where he proves, first, that there is a substantial interference with a right common to the public, as described above. Additionally, an individual claimant in public nuisance must prove that he has suffered special harm that is different from that burdening the public. To illustrate, it would not, for example, suffice for the individual claimant in the example above to bring an action in public nuisance claiming only that as a fisherman and a swimmer, defendant's conduct interfered with his ability to pursue those activities safely. To so claim, he would merely be stating the harm the community suffered, without any specific, individual, and *different* harm he himself suffered. On the other hand, what if our potential plaintiff owned the fishery with the contract to stock the lake with trout fingerlings on a periodic basis? If his contract with the town was canceled because the water became polluted, he would suffer the economic loss of that contract. He thereby would suffer an injury that was quite different from that suffered by the community at large. In this latter setting, our fishery owner might lodge a claim in public nuisance against the manufacturer even though the manufacturer's conduct encroached in no way upon fishery property or our plaintiff's right to enjoy his property. Similarly, as one court has stated, "in substantially all of those cases in which commercial fishermen using public waters have sought [nuisance] damages for the pollution or other tortious invasion of those waters, they have been permitted to recover[,]"[21] in that each can show "he has suffered a damage particular to him — that is, damage different in kind, rather than simply in degree, from that sustained by the public generally."[22]

It follows that an individual who has suffered a personal physical injury or illness as a result of a public nuisance may also be able to bring a public nuisance claim on behalf of himself. The rationale for so holding was stated by the court in *Anderson v. W. R. Grace & Co.*,[23] a claim by residents that defendant's introduction of toxic chemicals into groundwater caused severe personal injuries, including childhood leukemia. Finding plaintiffs could, as individuals, bring a claim in public nuisance, the court explained: "[W]hen a plaintiff has sustained 'special or peculiar damage,' he or she may maintain an individual action . . . Injuries to a person's health are by their nature 'special and particular,' and cannot properly be said to be common or public . . . As plaintiffs allege that they have suffered a variety of illnesses as a result of exposure to the contaminated water, they have standing to maintain this nuisance action.''[24]

In one suit claiming personal physical injury as well as property damage arising from defendant's contamination of groundwater, the court stated that plaintiffs claiming in public nuisance ''may only seek to obtain damages for their special injuries.''[25] Thus, the court added, if public nuisance was proved, plaintiffs would be able to recover for ''(1) the loss in rental value of their property, if any, (2) compensation for physical injuries, and (3) upon a showing of independent personal injury, damages for emotional distress.''[26]

4.5 TRESPASS

Trespass protects plaintiff's interest in the surface land itself, the earth or other material beneath the surface, and ''the air space above it.''[27] For this reason, depending upon the seriousness of the contamination of plaintiff's land or environment, defendant may be liable in trespass where the pollution or contamination interferes with plaintiff's possessory rights in the land, the land beneath it, or the ambient air.

The claim in trespass is readily confused with that for private nuisance. As explained by a leading authority, ''[t]he distinction which is now accepted is that trespass is an invasion of the plaintiff's interest in the exclusive possession of his land, while nuisance is an interference with his use and enjoyment of it.''[28] In most jurisdictions, invasions of plaintiff's property that amount to trespass may also, if they interfere with plaintiff's use and enjoyment of the property, be actionable in nuisance. In such circumstances, ''plaintiff may have his choice'' of a claim in trespass or in nuisance ''or may proceed upon both.''[29] It does not follow that a claim in nuisance ordinarily imports facts sufficient for a claim in trespass.

Plaintiff need not prove that defendant intended to trespass on the property. Rather, plaintiff must only prove that defendant intended the act that resulted in the trespass, i.e., that defendant's act was volitional, and done with knowledge to a substantial certainty that the act would result in introduction of the substance onto plaintiff's property.[30] For this reason, defendant may not defend an action in trespass by proving that he acted with the mistaken belief that his actions, and the consequent material invasion, were authorized by plaintiff. For example, in one suit, involving a fuel-oil distributor's mistaken fuel delivery to a residence and the resulting accidental spillage, the court rejected, as a defense, defendant's claim that it acted in a reasonable belief that the fuel had been ordered.[31]

A defendant may be liable in trespass even where the trespassory invasion causes no compensable harm to plaintiff's property or interest in the property.[32] The invasion of plaintiff's property need not be direct if plaintiff can prove that an intentional act of defendant resulted in the harm. Thus the causal intervention of natural conditions, such as deterioration, wind, or rain, in initiating or exacerbating the trespass will not absolve defendant of liability.[33] It was so held in one action where plaintiff claimed that defendant's dumping of asphalt waste on land contiguous to plaintiff's fish pond eventually resulted in the pollution of the pond.[34]

As in the doctrine of continuing nuisance, a polluter's failure to remove a pollutant or a contaminant from plaintiff's land may represent a "continuing" tort. The significance of the designation "continuing trespass" is primarily that of relieving some of the strictures of limitations periods within which the possessor would have to bring a toxic tort claim.[35]

NOTES

1. 238 N.C. 185, 77 S.E.2d 682 (1953).
2. *Id.*, 77 S.E.2d 682, citing Restatement, Second, Torts § 825.
3. Henderson and Pearson, *The Torts Process* 929 (3rd ed. 1988) ("When the interference is limited to light, noise, odor, or vibrations . . . courts generally hold that such intangible intrusions, in the absence of intent, are dealt with as nuisance cases, not trespass.").
4. *Bradley v. American Smelting and Refining Co.*, 104 Wash. 2d 677, 709 P.2d 782 (Wash. 1985). In that suit, involving intentional deposit of microscopic particles of arsenic and cadmium, the court added: "When, however, the particles or substance accumulat[e] on the land and d[o] not pass away, then a trespass has occurred." Id., 104 Wash. 2d at 691, 709 P.2d at 791.
5. Prosser, Wade and Schwartz, *Torts — Cases and Materials* 811 (8th ed.) (Foundation 1988). Restatement, Second, Torts § 821D, defines "private nuisance" as "a nontrespassory invasion of another's interest in the private use and enjoyment of land."
6. Henderson and Pearson, *The Torts Process* 929 (3rd ed. 1988), supra note 3.
7. *Baldwin v. McClendon,* 288 So. 2d 761 (Ala. 1976) (hog parlor proximate to plaintiff's land) (collecting authority).
8. For example, in *Jost v. Dairyland Power Cooperative,* 45 Wis. 2d 164, 172 N.W.2d 647 (1969), plaintiffs, crop growers with land proximate to defendant's coal burning electric generating plant, claimed that the plaint's daily discharge of approximately 90 tons of sulfur dioxide gas into the atmosphere whitened alfalfa leaves, killed or damaged pine and other trees, and generally lessened land values as a consequence of continuing crop loss. Rejecting appellee's argument that it should have been permitted to introduce evidence of the social and economic utility of its generating plant, the Wisconsin Supreme Court stated: "Whether its economic or social importance dwarfed the claim of a small farmer is of no consequence in this lawsuit. It will not be said that, because a great and socially useful enterprise will be liable in damages, an injury small by comparison will go unredressed . . . To contend that a public utility, in pursuit of its praiseworthy and legitimate enterprise, can, in effect,

deprive others of the full use of their property without compensation, poses a theory unknown to the law of Wisconsin, and in our opinion would constitute the taking of property without due process of law . . . " *Id.*, 172 N.W.2d at 653–654.

9. An example of comparative injury analysis is found in *Crushed Stone v. Moore*, 369 P.2d 811 (Okla. 1962). There the court stated: "While we recognize that in proper cases, especially involving businesses upon which the public's interest, or necessity, depends, the matter of 'comparative injury' should be given prominent consideration, this court is among those holding that where damages in an action at law will not give plaintiffs an adequate remedy against a business operated in such a way that it has become a nuisance, and such operation causes plaintiffs substantial and irremediable injury, they are entitled, as a matter of right, to have same abated, by injunction, ' . . . notwithstanding the comparative benefits conferred thereby or the comparative injury resulting therefrom.' [citation omitted]." *Id.*, 369 P.2d at 815–816.
10. *Vickage 1st and 2nd Addition Homeowners Ass'n v. Catholic Diocese*, 510 P.2d 1296 (Kan. 1976).
11. *Id.*, 510 P.2d 1296.
12. E.g., *Crushed Stone Co. v. Moore*, 369 P.2d 811 (Okla. 1962) (the injury claimed by plaintiff included, inter alia, aggravation of allergies and worsening of one resident's nervous condition).
13. Restatement, Second, Torts § 840A ("Continuing Liability After Transfer of Land.").
14. 762 F.2d 303 (3d Cir. 1985).
15. The appeals court states: "We believe that this result is consonant with the historical role of private nuisance law as a means of efficiently resolving conflicts between *neighboring* contemporaneous land uses. . . . Neighbors, unlike purchasers of the land upon which the nuisance exists, have no opportunity to protect themselves through inspection and negotiation." *Id.*, 762 F.2d at 314.
16. *Id.*
17. Restatement, Second, Torts 821B.
18. *Philadelphia Electric Co. v. Hercules, Inc.*, 762 F.2d 303 (3d Cir. 1985).
19. 370 F. Supp. 247 (D. Me. 1973).
20. *Id.*, 370 F. Supp. at 249–250.
21. *Id.*, 370 F. Supp. at 250.
22. *Id.*, 370 F. Supp. at 250.
23. 628 F. Supp. 1219, 1233 (D. Mass. 1986).
24. *Id.*, 628 F. Supp. at 1233.
25. *Anderson v. W. R. Grace & Co.*, 628 F. Supp. 1219, 1233 (D. Mass. 1986).
26. *Id.*, 628 F. Supp. 1219, 1233 (D. Mass. 1986).
27. Restatement, Second, Torts § 158 comment i.
28. *Prosser and Keeton on the Law of Torts* 622 (5th ed. 1984).
29. *Mangini v. Aerojet-General Corp.*, 281 Cal. Rptr. 827, 230 Cal. App. 3d 1125 (Cal. App. 1991) (quoting Restatement, Second, Torts § 821D, comment e) (plaintiff's land contamination claim in nuisance, trespass, and other common law causes of action).
30. See Restatement, Second, Torts § 158 comment i: "It is enough that an act is done with knowledge that it will to a substantial certainty result in the entry of a foreign matter."

31. *Serota v. M. & M. Utilities, Inc.*, 55 Misc. 2d 286, 285 N.Y.S.2d 121 (1967) ("Obviously, the defendant intended to come upon plaintiff's land and make an oil delivery and did not intend to commit a trespass or intentionally to cast oil upon plaintiff's land. His innocence and his mistaken belief that his visit was authorized is of no moment since his intent is clearly shown to have been to deliver oil. This unauthorized act, resulting in whatever damages which may have occurred, rendered him liable." *Id.* at 124.).
32. One may be "subject to another for trespass, irrespective of whether he thereby causes harm to any legally protected interest of the other[.]" Restatement, Second, Torts § 158.
33. Comment i to Restatement, Second, Torts § 158 gives these examples: "[O]ne who piles sand so close to his boundary that by force of gravity alone it slides down on to his neighbor's land, or who builds an embankment that during ordinary rainfalls the dirt is washed from it upon adjacent lands, becomes a trespasser on the other's land."
34. *Rushing v. Hooper-McDonald, Inc.*, 300 So. 2d 94 (Ala. 1974) ("This court holds that it is not necessary that the asphalt or foreign matter be thrown or dumped directly and immediately upon the plaintiff's land but that it is sufficient if the act is done so that it will to a substantial certainty result in the entry of the asphalt or foreign matter onto the real property that the plaintiff possesses.").
35. Restatement, Second, Torts § 161, comment b.

CHAPTER 5

Negligence

CONTENTS

Negligence

5.1 GENERALLY

In negligence, plaintiff must prove that defendant failed to act with due care under the circumstances and that this act or omission to act proximately caused plaintiff's injury or loss. Negligence liability may follow a manufacturer's failure to exercise due care in the: (1) design or formulation of the product; (2) selection of proper materials for the product; (3) manufacture, fabrication, or packaging of the product; or (4) provision of warnings or instructions for reasonably safe use of the product.

As stated by one court considering the appeal of a claim of injurious exposure to fumes emanating from a container of potato whitener, "[i]n an action to recover for personal injuries resulting from [a] manufacturer's negligence, plaintiff must produce evidence which tends to show that the product manufactured was defective at the time it left the defendant-manufacturer's plant, and that the defendant-manufacturer was negligent in its design of the product, in its selection of materials, in its assembly process, [or] in inspection of the product."[1]

The manufacturer also has a duty to inform those likely to be exposed to a potentially hazardous product of the risks of such exposure and the precautionary measures, if any, that will permit persons to use the product in a reasonably safe manner. As the court in the leading asbestos exposure case of *Borel v. Fibreboard Paper Products Corp.*[2] stated: "[a] product must not be made available to the public without disclosure of those dangers that the application of reasonable foresight would reveal."[3] This duty to warn, in both negligence and strict products liability,[4] is nondelegable. The *Borel* court also stated: "[A] manufacturer [may not] rely unquestioningly on others to sound the hue and cry concerning the danger of its product. Rather, each manufacturer must bear the burden of showing that its own conduct was proportionate to the scope of its duty."[5]

5.2 NEGLIGENCE DISTINGUISHED FROM OTHER CLAIMS

As in actions for warranty and strict tort liability, plaintiff must produce evidence that the product left the manufacturer's control in the defective or unreasonably dangerous condition.[6] Nevertheless, plaintiff's burden of proof in negligence differs from the proof required in various strict liability toxic harm remedies, such as nuisance, trespass, strict tort liability, warranty, and liability for abnormally dangerous activities. In the latter remedies, plaintiff's greatest burden is proving the nature and injurious consequences of defendant's conduct, but plaintiff does not have to show that defendant's acts breached a duty of ordinary care under the circumstance.[7]

In some jurisdictions, the defendant may be found to have breached its implied warranty of merchantability without having been negligent, but cannot be found negligent without a finding that the product or substance also breached the implied warranty of merchantability. In *Caccavale v. Raymark Industries, Inc.*,[8] a wrongful death action brought on behalf of an asbestos worker against a manufacturer of asbestos products, the Massachusetts Supreme Judicial Court reversed a trial court jury verdict finding that defendant was negligent, but that its products had not breached the implied warranty of merchantability.

Suppose, on the other hand, that a manufacturer sold a potentially toxic product that was later implicated in the injury to a person or his property, and suppose further the evidence was sufficient to persuade the jury that at the time of initial manufacture and sale the manufacturer acted with ordinary care under the circumstances and that the product also was fit for its ordinary purpose, i.e., that it did not breach the implied warranty of merchantability. If, after initial manufacture and sale, it became known to the industry that the product had to be recalled to avoid an unreasonable risk of harm, or repaired in order to permit it to be used safely, in many jurisdictions the manufacturer may have a post-sale duty to warn of product risks.[9] In such a setting, a manufacturer might be found liable in negligence if it breached its post-sale duties to purchasers, even though it would not be found liable for breach of the implied warranty of merchantability at the time of initial sale.[10]

5.3 THE ACTOR'S DUTY OF ORDINARY CARE UNDER THE CIRCUMSTANCES

In most circumstances, the toxic tort plaintiff can show that defendant owed her a duty of due or ordinary care under the circumstances. In general terms, defendant owes plaintiff a duty of ordinary care where plaintiff and defendant are so circumstanced that a reasonable person in defendant's position would recognize that acting without due care would create an unreasonable risk of injury or loss to plaintiff.

Applying this standard to the toxic torts context, the manufacturer of hazardous substances, or those engaged in their transportation, storage, application, or like pursuits, can readily foresee that failure to exercise due care in formulation, production, packaging, transportation, labeling, or related pursuits will pose a risk of harm to workers, users, or bystanders. They are in a position to clearly apprehend certain foreseeable risks, hence their duty to take appropriate precautions. Thus, absent extraordinary circumstances, the agricultural worker administering a pesticide manufactured by defendant, the industrial worker handling toxic solvents, or the landowner adjoining a smelting plant can easily show defendant's duty.[11]

An actor's duty of care is commensurate with the risks posed by exposure to the product or substance. Restatement, Second, Torts § 298 defines reasonable care as "the care . . . which a reasonable man [in the actor's position] would recognize as necessary to prevent the act from creating an unreasonable risk of harm to another." Hence, where the risk of harm is great, or where the injury that may occur is serious, even if the risk of it occurring is relatively small, the care and vigilence required of the manufacturer is raised accordingly.

A product manufacturer is expected to employ testing and inspection procedures with a thoughtfulness and thoroughness commensurate with those employed by prudent members of the same manufacturing community at the time the challenged product was manufactured and sold.[12] The duty to test and inspect extends to packaging or containers of a potentially hazardous substance. A high degree of care is imposed where a flaw in packaging or containment of a product or substance creates a foreseeable hazard to human life.[13] The absence of dispositive testing or screening procedures at the time a product is distributed will not absolve the provider of liability if it can be shown that a prudent provider would have utilized testing or screening means then available.[14]

However phrased, the manufacturer owes the user or consumer adequate warnings where the product poses an unreasonable risk of injury and instructions adequate to alert the ordinary user of means to avoid the risk or reduce the risk to acceptable levels.[15] In negligence law, an exception to the manufacturer's general duty to warn of foreseeable product risks is where it is shown that those who will be exposed to the product are experienced professionals who can be presumed to be familiar with pertinent risks. For example, in *Martinez v. Dixie Carriers, Inc.*,[16] the manufacturer of a petrochemical mixture containing concentrations of benzene was not liable for failure to warn of the risks of inhalation by experienced tank strippers, due to limited marketing of the product and general awareness (within the industry) of the pertinent risks.[17]

5.4 BREACH OF DUTY

In negligence, plaintiff must show that defendant breached a duty of due care owed plaintiff. The toxic tort plaintiff must, therefore, introduce evidence of what ordinary care under the circumstances would have required of defendant, and how defendant's conduct fell below that standard.

A manufacturer is in breach of its duty of care to another where the facts show that the burden of taking measures that would have lessened the risk of harm was less than the burden (on plaintiff and on society more generally) of the risk or hazard created by the product (borne by plaintiff or the population generally). The latter burden, i.e., the one placed upon plaintiff and others by the manufacturer's creation of the risk, is often described as the multiple of (P) [the probability that, without the manufacturer's precautionary or remedial measures, a person will be injured by use of or exposure to a product or substance] and (L) [the magnitude of the injury should it occur].

The burden on the manufacturer (B) of making the product reasonably safe might, for example, entail its changing the design or formulation of the product, changing its container or means of dispensing, or adding warnings or instructions as to its safe use. A manufacturer is considered to have *breached* its duty of care to another where the burden upon it of precautionary measures (B) is less than the probability that harm may occur (P) multiplied by the magnitude of the injury should injury occur at all (L). Thus it is stated often that a manufacturer has breached its duty where (B) is less than (P) (L), or (B) < (P) (L).[18]

In one illustrative suit, the asbestos case of *Graham v. Pittsburgh Corning Corp.*,[19] the trial court, considering plaintiff's claim that the manufacturer breached its duty to warn, described "a test which balances the severity of the harm which might result from use of the product against the burden or cost of protective action by the manufacturer."[20] The standard of care expected of a manufacturer at the time it introduced a product into the marketplace may be "established by looking to the conduct of the industry or profession in similar circumstances as of that date.[21]

Evaluation of an actor's conduct may take into account pertinent statutes or regulations. There may, for example, be the Toxic Substance Control Act[22] or Federal Insecticide, Fungicide, and Rodenticide Act[23] standards that apply to the labeling of a product, or there may be federal environmental statutes that govern the manufacturing or waste disposal practices of an industry. An actor's ability to show that his conduct satisfied the pertinent requirements of a statute or a regulation may absolve them of negligence liability where the court concludes either that (1) the statute or regulation represents a standard of due care under the circumstances or (2) where the statute or regulation expressly or impliedly preempts application of the negligence law of a particular state.[24]

With regard to some claims, on the other hand, an actor's conformity with the requirements of a statute or a regulation may only be considered evidence that defendant acted with due care. Where the latter approach is adopted, the court or the jury is empowered to decide that defendant should reasonably be expected to have done more to reduce or eliminate the risk. For example, in *Jonescue v. Jewell Home Shopping Service*,[25] involving a child's ingestion of a cleaning fluid, the seller's showing that the fluid was considered nontoxic under the Federal Hazardous Substances Labeling Act did not preclude a showing that the product's container and coloration made the product's hazards unobvious.

5.5 DAMAGES — THE ECONOMIC-LOSS RULE

Often the toxic tort plaintiff has not suffered either personal physical injury or damage to property. Rather, the loss suffered may have been confined to economic loss, i.e., loss of revenues or business occasioned by a product's failure to perform as expected or a plaintiff's inability to conduct his business profitably for a period of time. In these circumstances, courts in many jurisdictions find that plaintiff's claim for "economic loss" not accompanied by other physical injury or damage cannot be redressed in tort. Instead, this doctrine provides, plaintiff's claim for exclusively economic loss must proceed, if at all, in a claim for breach of implied or express warranty.

The pesticide-related suit in *Prairie Production, Inc. v. Agchem Div.-Pennwalt Corp.*[26] illustrates this rule. In that case, plaintiff's negligence and implied warranty action claimed that a manufacturer's pesticide, Penncap-M,® was ineffective in combatting corn earworms, causing a loss of a proportion of plaintiff's seed-corn crop to the pest. While suggesting that plaintiffs might, upon certain proof, pursue the claim for economic loss through their warranty claim, the Indiana appellate court applied the general rule that "a manufacturer does not owe a duty to avoid causing purely economic loss."[27] As its rationale, the court explained that the warranty provisions of the Uniform Commercial Code "gover[n] the economic relations between buyer and seller, and the dissatisfied buyer may avail himself of the remedies fashioned by the legislature. . . . Also, relegating the purchaser to warranty remedies prevents the manufacturer from being held liable for damages of unknown and unlimited scope."[28]

NOTES

1. *Sutton v. Major Products Co.*, 91 N.C. App. 610, 8 U.C.C. Rep. Serv. 2d 376, 372 S.E.2d 897, 898 (N.C. App. 1988).
2. 493 F.2d 1076 (5th Cir. 1973).
3. *Id.*, 493 F.2d 1076 at 1090.
4. Products liability without fault, or strict products liability, is discussed in Chapter 7. A manufacturer's or seller's duty to warn is discussed in Chapter 9.
5. *Id.*, 493 F.2d 1076, 1090.
6. *Sutton v. Major Products Co.*, 91 N.C. App. 610, 8 U.C.C. Rep. Serv. 2d 376, 372 S.E.2d 897 (N.C. App. 1988) (summary judgment granted against grocery store worker's action against manufacturer of potato whitener claimed

to emit noxious fumes, the court explaining: "Because plaintiff failed to forecast evidence of a defect in the product in existence at the time the product left Major Products' plant, summary judgment in favor of Major Products was properly allowed in the negligence claim." *Id.*, 372 S.E.2d at 898.

7. Thus in warranty plaintiff must show the unmerchantable nature of the product; in strict tort liability, the defective and unreasonably dangerous condition of the product; in nuisance, the unreasonable interference with plaintiff's use and enjoyment of real property; etc. In virtually all settings, defendant's motivation and prudence, or lack of prudence, is immaterial.
8. 404 Mass. 93, 533 N.E.2d 1345, Prod. Liab. Rep. (CCH) ¶ 12,108 (Mass. 1989).
9. The post-sale duty to warn is described in Chapter 9.
10. See *Caccavale v. Raymark Industries, Inc.*, 404 Mass. 93, 533 N.E.2d 1345, Prod. Liab. Rep. (CCH) ¶ 12,108 (Mass. 1989), discussing *Hayes v. Ariens Co.*, 391 Mass. 407, 410, Prod. Liab. Rep. (CCH) ¶ 10,030 (Mass. 1984).
11. Only where the circumstances of plaintiff's contact with the toxin is quite bizarre, remote, or extreme, which is to say, not reasonably foreseeable, will defendant claim successfully that it owed plaintiff no duty.
12. Cf., *McKee v. Miles Laboratories, Inc.*, 675 F. Supp. 1060, Prod. Liab. Rep. (CCH) ¶ 11,733 (E.D. Ky. 1987) (manufacturer of blood coagulant).
13. See generally *Vuono v. New York Blood Center, Inc.*, 696 F. Supp. 743, Prod. Liab. Rep. (CCH) ¶ 11,935 (D. Mass. 1988) (claim arising from packaging blood products in an unsterile vial).
14. E.g., *Doe v. Univ. Hosp of N.Y. Univ. Hosp. Center,* 148 Misc. 2d 756, 561 N.Y.S.2d 326 (N.Y. Sup. 1990) (prior to development of conclusive test for blood contamination by the AIDS virus, blood provider could be liable in negligence for failure to utilize available safeguards).
15. E.g., *Graham v. Pittsburgh Corning Corp.*, 593 A.2d 567 (Del. Sup. 1990) ("*Prosser and Keeton on Torts* (W. Keeton 5th ed. 1984 § 96 and 1 Madden, *Products Liability* (2nd ed. 1988) § 10.1 require warning where the product poses unreasonable risk to the user." *Id.* at 569).
16. 529 F.2d 457, Prod. Liab. Rep. (CCH) ¶ 7643 (5th Cir. 1976).
17. Seller informational duties are discussed in Chapter 9.
18. *Cf.*, *Graham v. Pittsburgh Corning Corp.*, 593 A.2d 567 (Del. Super. 1990) (citing with approval a test for breach "which balances the severity of harm which might result against the burden of protective action by the manufacturer." and 1 Madden, *Products Liability* 2nd ed. 1988 § 4.2).
19. 593 A.2d 567 (Del. Super. 1990).
20. 593 A.2d 567, 569 (Del. Super. 1990).
21. See, e.g., *McKee v. Miles Laboratories, Inc.*, 675 F. Supp. 1060, Prod. Liab. Rep. (CCH) ¶ 11,733 (E.D. Ky. 1987) (manufacturer of blood coagulant Factor VIII).
22. 15 U.S.C. § 2601 *et seq.*
23. 7 U.S.C. § 136 *et seq.*
24. The doctrines of express and implied preemption are discussed in Chapter 14.
25. 306 N.E.2d 312, Prod. Liab. Rep. (CCH) ¶ 7121 (Ill. App. 1974).

26. 514 N.E.2d 1299, Prod. Liab. Rep. (CCH) ¶ 11,640 (Ind. App. 1987).
27. *Id.*, Prod. Liab. Rep. (CCH) ¶ 11,640 at 33,105.
28. *Id.*, Prod. Liab. Rep. ¶ 11,640 at 33,1105 (citations omitted).

CHAPTER 6

Warranty

CONTENTS

Warranty

6.1 GENERALLY

A "warranty" is a seller's representation, be it expressed or implicit, as to a *material* quality of a product. Generally, warranty obligations attach only to persons who sell such products in the ordinary course of their business. The distinction between express (explicit) and implied (implicit) warranties is this: an express warranty is a seller's *stated* affirmation of a product's performance, efficacy, or safety.[1] To be contrasted, the sale of any product by a merchant of such products carries with it an *implied* warranty that the product is fit for the ordinary purposes for which it is sold. An additional type of implied warranty, the implied warranty of fitness for a particular purpose, is created where a buyer approaches a seller with a request for a product of particular and specialized qualities. It arises where a seller with superior knowledge of a product's qualities, performance, or safety sells that product to one whom the seller knows is relying upon him for a measure of expert advice as to the suitability of the product for the buyer's needs. This warranty may be created even by a seller who is not ordinarily a merchant of such goods.

6.2 EXPRESS WARRANTIES

Product sellers may create express warranties when they make affirmative claims about the quality, safety, performance, and durability of their product. The seller's affirmation as to safety or performance may be spoken, written, broadcast, or any other conventional form of communication. Plaintiff need not show that he relied upon the warranty in deciding to purchase the product; if they can produce evidence of an express warranty of performance or safety, there is a presumption that they relied upon the warranty in making the purchase.[2]

Pertinent to toxic tort claims, a product that fails to perform as safely or as effectively as the seller claimed can create a claim in breach of express warranty. For example, a pesticide seller's claims that its product may be safely administered by a person not using respiratory protection would breach that express warranty where the advised use caused the user a respiratory ailment. The seller need not use the word "safe" to create an express warranty as to safety. Illustrative is the Tennessee appellate decision in *Perma-Strate Co. v. Gemus,*[3] involving scalp injuries suffered by the user of a hair straightener bearing a label that announced: "Absolutely will not burn or discolor normal hair. Will not irritate scalp when used as directed.'' A claim that a product offers ''virtually'' complete effectiveness may, however, not support the conclusion that the seller warranted ''absolute'' effectiveness.[4]

A variety of express warranty claims arise concerning the sale of agricultural chemicals. Regarding breach of an express warranty as to performance, an agricultural fungicide advertized to retard fungus damage to, for example, potatoes, would breach that express warranty if it failed to have that retardant effect. The fungicide would also breach the express warranty if, even while operating effectively as a fungicide, it's prescribed application contaminated the soil so as to delay or preclude planting. For example, in *Swenson v. Chevron Chemical Co.*,[5] the South Dakota Supreme Court held that a label listing chemicals contained in defendant's insecticide constituted an express warranty that the product would perform effectively as an insecticide.

Spoken assurances, including a salesman's representations, may create an express warranty. For example, a federal appeals court permitted plaintiff dairy farmers to recover actual and punitive damages against the manufacturer of a cattle feed where plaintiffs' purchases were based upon a salesman's commendation of the product's performance. The additive, Proto-Tone,® while promoting a temporary increase in herd milk production, caused illness in the cows with a consequent substantial loss in milk production.[6] Similarly, in *Venie v. South Central Enterprises*,[7] the manufacturer's agent's statements to the buyer that defendant's 2,4,5-T, a hormone herbicide, was ''perfectly safe for strawberries'' was actionable in breach of express warranty when, following application of the product in a normal manner, it killed not only the target weeds, but the strawberries as well.

6.3 IMPLIED WARRANTY OF MERCHANTABILITY

Uniform Commercial Code § 2-314 provides that to be ''merchantable,'' a product must, at the least, ''(a) pass without objection in the trade under the contract description; and (b) in the case of fungible goods, [be] of fair average quality within the description; and (c) [be] fit for the ordinary purpose for which such goods are used; and (d) run, within the ordinary variations of the agreement, of even kind, quality and quantity within each unit and among all units involved; and (e) [be] adequately contained, packaged and labeled as the agreement may require; and (f) conform to the promises or affirmations of fact made on the container or label if any.''

While most toxic tort claims against cigarette manufacturers have failed to prove liability in negligence or strict products liability, one decision held that plaintiffs' claims that cigarettes breached the implied warranty of merchantability should go to trial. Denying defendants' motion for summary judgment, the court agreed that plaintiffs' claim that the manufacturer impliedly warranted that its products are ''merchantable, fit for [their] intended use, free from defect, safe to consume and non-carcinogenic and non-addictive'' successfully stated a claim in "breach of implied warranty based upon an alleged design defect."[8]

Distinctive treatment has been accorded pharmaceuticals and medical-biological products that are characterized as "unavoidably unsafe" within the meaning of comment k, Restatement, Second, Torts § 402A. There is authority providing that granting "unavoidably unsafe" status to a defendant blood supplier's product, contaminated with the AIDS virus and provided to plaintiff prior to the availability of a feasible method of screening blood for this contamination, precluded plaintiff's additional claims in breach of the implied warranty of merchantability and breach of the implied warranty of fitness for a particular purpose.[9]

Significantly, some courts have preserved plaintiffs' implied warranty of merchantability claims against toxic tort medical device defendants even while finding plaintiffs' failure to warn claims preempted by federal law.[10] For example, in *Krause v. Kimberly-Clark Corp.*,[11] the federal trial court granted summary judgment against plaintiff's toxic shock claims against defendant manufacturer of tampons, but held that plaintiff's negligence and implied warranty claims raised triable issues.[12]

The principal contribution of the implied warranty claim upon toxic tort litigation is this: the toxic tort plaintiff often joins an implied warranty claim with tort claims in negligence and strict products liability. It is agreed generally that if product is unreasonably dangerous for its intended or reasonably foreseeable use it is not, by definition, fit for its ordinary purpose. Thus, where a plaintiff successfully proves that the product was sold in an unreasonably dangerous and defective condition and that the product's toxicity proximately caused personal injury or property damage, plaintiff has not only proven negligence and strict products liability, but has *also* almost certainly proved breach of the implied warranty of merchantability.

6.4 TO WHOM WARRANTIES RUN

A remedy in breach of express or implied warranty may be available against the manufacturer, to any person who has actually purchased the product, either directly or through an intermediate seller, such as a wholesaler or a retail seller.[13] There exist substantial state-by-state variations in the designation of persons *other than the purchaser* who may being claims in express or implied warranties. By the adoption of one of the three variations set forth in Uniform Commercial Code § 2-318 Alternatives A, B, and C, state legislatures have determined the classes of persons who may bring such claims. These alternatives read:

Alternative A

> A seller's warranty whether express or implied extends to any natural person who is in the family or household of his buyer or who is a guest in his home if it is reasonable to expect that such person may use, consume, or be affected and who is injured in person by the goods. A seller may not exclude or limit the operation of this section.

Alternative B

> A seller's warranty whether express or implied extends to any natural person who may reasonably be expected to use, consume, or be affected by the goods and who is injured in person by the breach of the warranty. A seller may not exclude or limit the operation of this warranty.

Alternative C

> A seller's warranty whether express or implied extends to any person who may reasonably be expected to use, consume, or be affected by the goods and who is injured by the breach of the warranty. A seller may not exclude or limit the operation of this section with respect to injury of the person of an individual to whom the warranty extends.

Alternative A has been adopted by the largest number of states,[14] and is the most restrictive application of the rule, permitting a claim only by (a) a natural person (excluding business entities) and (b) a family or household member, or a house guest. Only personal, physical injury is redressable.

Alternative B extends the class of third-party beneficiaries beyond family members and household guests to include natural persons who might be reasonably foreseen to use or be affected by the product. Significant to many toxic tort claims, this designation would include employees and bystanders.

Alternative C, the most expansive version a state may adopt, extends warranty protection to (a) businesses, in addition to natural persons, and (b) injuries other than injuries to the person, including property damage and economic loss. The scope of remedies Alternative C offers to nonpurchasers compares with that provided in the strict products liability remedy of Restatement, Second, Torts § 402A.[15]

6.5 WARRANTY DISCLAIMERS AND LIMITATIONS

6.5.1 Disclaimers

Implied warranties of merchantability and of fitness for a particular purpose may be disclaimed by properly worded and conspicuous expression in the selling documents. UCC § 2-316(2) states guidelines for the disclaimer of implied warranties, and generally provides that a seller may disclaim the implied warranty of merchantability by a "conspicuous" expression that mentions "merchantability" specifically. "Language to exclude all implied warranties of fitness," the section continues, "is sufficient if it states, for example, that 'There are no warranties which extend beyond the description on the face hereof.'" An "as is" sale may operate to exclude all implied warranties; § 2-316(3)(a) states and explains that "unless the circumstances indicate otherwise, all implied warranties are excluded by expressions like 'as is,' 'with all faults', or other language which in common understanding calls the buyer's attention to the exclusion of warranties and makes plain that there is no implied warranty[.]"

6.5.2 Limitations

By explicit provision in sales documents, a seller may "limit or alter" a buyer's measure of damages. UCC § 2-719(1)(a) provides the example that such a limitation might confine the "buyer's remedies to return of the goods and repayment of the [purchase] price[.]" A seller may, by warranty, limit or exclude consequential damages unless the limitation or exclusion is unconscionable.[16] The Code states that "[l]imitation of consequential damages for injury to the person is prima facie unconscionable[.]"[17]

Even absent personal injury, a warrantor's remedy limitations may be found unconscionable in a setting where, in the buyer's particular circumstances, recognition of the limitation would strip the buyer of any reasonable warranty protection. Such unconscionability has been found where a claimed defect in a herbicide caused the failure of most of plaintiff's corn crop.[18]

NOTES

1. E.g., *Swenson v. Chevron Chemical Co.*, Prod. Liab. Rep. (CCH) ¶ 7525 (S.D. 1975) (insecticide bag's listing of chemicals for control of rootworm larvae constituted express warranty of insecticide's effectiveness).
2. U.C.C. § 2-313, comment 3 states: "In actual practice affirmations of fact made by the seller about the goods during a bargain are regarded as part of the description of those goods; hence no particular reliance on such statements need be shown in order to weave them into the fabric of the agreement. Rather, any fact which is to take such affirmations, once made, out of the agreement requires clear affirmative proof."
3. Prod. Liab. Rep. (CCH) ¶ 5885 (Tenn. App. 1967).
4. Cf., *Whittington v. Eli Lilly & Co.*, 333 F. Supp. 98, Prod. Liab. Rep. (CCH) ¶ 6679 (D.W. Va. 1971) (plaintiff who became pregnant after using birth control pill manufactured by defendant).

5. Prod. Liab. Rep. (CCH) ¶ 7525 (S.D. 1975).
6. *Boehm v. Fox,* 473 F.2d 445, Prod. Liab. Rep. (CCH) ¶ 6984 (10th Cir. 1973).
7. 401 S.W.2d 495, Prod. Liab. Rep. (CCH) ¶ 5531 (Mo. App. 1966).
8. *Kyte v. Phillip Morris, Inc.,* 408 Mass. 162, 556 N.E.2d 1025, Prod. Liab. Rep. (CCH) ¶ 12,607 at 37,694 (Mass. 1990) ("The plaintiffs state that their claim is not based on the proposition that all cigarettes are inherently defective, but rather on a claim that, because of their defective design, [defendant's] cigarettes were inherently carcinogenic and addictive. We accept their characterization of their own claim which is consistent with the allegations of the breach of implied warranty count." *Id.* at 37,694.).
9. *Miles Laboratories v. Doe,* 315 Md. 704, 556 A.2d 1107, Prod. Liab. Rep. (CCH) ¶ 12,126 (Md. App. 1989). In its response to a question certified by a federal trial court, the Maryland court stated: "[W]e conclude that implied warranty claims for personal injury, consequential, or wrongful death damages cannot be sustained where the claim for strict tort liability under 402A falls under Comment k — that when, because of an unknown and unknowable virus contained in the blood product which is undetectable by any available scientific test, the product is incapable of being made entirely safe, yet must be marketed due to the profound and essential public need for it." *Id.* at 35,333. The infection at issue in the case occurred in 1983. *Id.* at 35,328.
10. A discussion of express and implied preemption may be found in Chapter 14.
11. 749 F. Supp. 164, Prod. Liab. Rep. (CCH) ¶ 12,511 (W.D. Mich. 1990).
12. The court stated: "Consistent with its determination that only [the] state law clai[m] which challenges the adequacy of information provided with tampon packages is preempted, the Court concludes that plaintiff's claims of negligence and breach of implied warranty are not preempted. Based on the apparent lack of medical certainty available with regard to [Toxic Shock Syndrome], the court notes that plaintiff may have a difficult time establishing her negligence and breach of warranty claims." *Id.*, ¶ 12,511 at 37,195.
13. See generally *Henningsen v. Bloomfield Motors, Inc.*, 32 N.J. 358, 161 A.2d 69 (1960) (personal injury suit brought by wife of purchaser of vehicle; held: wife's warranty claims could be brought against both the manufacturer and the dealer).
14. Alaska, Arizona, Connecticut, District of Columbia, Florida, Georgia, Idaho, Illinois, Indiana, Kentucky, Michigan, Mississippi, Missouri, Montana, Nebraska, New Jersey, New Mexico, North Carolina, Ohio, Oklahoma, Oregon, Pennsylvania, Tennessee, the Virgin Islands, Washington, West Virginia, Wisconsin.
15. Strict tort liability as defined in Restatement, Second, Torts § 402A is discussed in Chapter 7.
16. E.g., U.C.C. 2-719(3); S.D. Codified Laws Ann. § 57A-2-719(3) (1980).
17. UCC § 2-719(3).
18. *Herrick v. Monsanto Co.*, 834 F.2d 594, Prod. Liab. Rep. (CCH) ¶ 12,142 at 35,408 (8th Cir. 1989) (imposition of the limitation of consequential damages would leave plaintiff without an adequate remedy where crop loss is inevitable where the herbicide is ineffective).

CHAPTER 7

Strict Tort Liability

CONTENTS

Strict Tort Liability

7.1 GENERALLY

7.1.1 Rationale for Strict Products Liability

The complexity of modern research, development, and manufacturing processes makes it quite difficult for plaintiff to prove that a defendant was negligent in making or selling a potentially hazardous substance. Even with the most capable counsel and expert witnesses, it is very difficult, and often impossible, for plaintiff to identify and present evidence that defendant acted without due care under the circumstances. Conversely, the manufacturer or marketer of a potentially deleterious substance can, with far greater ease, offer proof that the procedures and processes employed were both cautious and prudent under the circumstances.

The rationale for adopting strict products liability was to relieve plaintiff of the need to show that defendant acted with an absence of due care. In strict products liability, plaintiff need not show that defendant's conduct was deficient in some way, or that defendant acted without due care under the circumstances, i.e., that defendant was negligent. Rather, plaintiff need only prove the consequences of defendant's action: that defendant sold a product in a defective and unreasonably dangerous condition and that plaintiff was injured as a foreseeable consequence of the product's hazardous condition.

Strict liability for the sale or marketing of unreasonably dangerous products is a fairly recent theory of legal remedy. It arose during the 1960s with its endorsement by the California Supreme Court in the influential decision of *Greenman v. Yuba Power Products, Inc.*[1] and its adoption, shortly thereafter, as Restatement, Second, Torts § 402A.

7.2 ELEMENTS

To recover in strict products liability, plaintiff must show (1) that defendant manufactured, sold, or marketed the product; (2) that the product was in an unreasonably dangerous, defective condition; (3) that plaintiff purchased or was exposed to the product when it was in substantially the same condition as it was upon leaving the control of defendant; and (4) that plaintiff suffered injury to their person or to their property due to the defective condition of the product.[2] The prevalent formulation of strict tort liability is that found in Restatement, Second, Torts § 402A. That section provides that a product seller shall be liable without regard to fault (or due care) where he sells a product that is in a "defective, unreasonably dangerous condition" and the product defect proximately causes bodily injury or property damage.[3]

In products liability actions, the toxic tort claimant must prove that the claimed injury or loss resulted from a condition in the product that rendered it unreasonably dangerous[4] and that the hazardous condition existed at the time the product left the manufacturer's control.[5] Some jurisdictions have eliminated the requirement that plaintiff prove the "unreasonably dangerous" condition of the product, ruling that the claimant must only demonstrate a defective condition and consequent, foreseeable injury to plaintiff. However, in most jurisdictions, courts require of claimant a demonstration that the product was sold in an unreasonably dangerous condition.[6]

7.2.1 Negligence Distinguished

In a negligence claim against the seller of a toxic product, plaintiff must prove that the seller acted without due care under the circumstances. As related in the discussion in Chapter 5, the manufacturer is held to the standard of an expert in the pertinent field, and if a reasonable manufacturer *should* have known of a toxic risk and failed to remedy it or provide sufficient warnings, plaintiff's resulting injury may be redressable in negligence.

In the strict products liability claim, on the other hand, the gist of plaintiff's claim is the defective and unreasonably dangerous condition of the product, not the conduct of the manufacturer. Irrespective of whether the manufacturer knew or should have known of a product's risks, if these risks were scientifically knowable at the time of manufacture, the manufacturer may be strictly liable for resultant personal injury or property damage. In its distinctions from the negligence claim, the strict liability action resembles negligence in this way: if risks posed by a toxic product were known or knowable among experts in the field, that knowledge is imputed to the manufacturer. On the basis of this imputed knowledge, the manufacturer would have been negligent in the manufacture or marketing of the product, and if injury results, the manufacturer will be strictly liable for the harm caused.

There generally is agreement that "weighing somewhat more heavily in favor of recovery in a strict products liability action than in ordinary negligence actions is the policy of diverting the burden of product-caused injuries from the innocent victims to the manufacturer."[7]

7.3 UNREASONABLY DANGEROUS, DEFECTIVE CONDITION

7.3.1 Generally

Two primary means are employed in deciding if a product or a substance is in an unreasonably dangerous, defective condition. One, the "consumer expectations" test, measures the degree of danger by reference to what degree of risk would be expected by a consumer of ordinary knowledge and perception. The second, a "risk-utility" analysis, invites evaluation of a variety of factors ranging from a consumer's expectations to the availability of a feasible, commercially practicable product that would reduce or eliminate the challenged risk.

7.3.2 Consumer Expectation Test

The "consumer expectations" test, referred to in the Reporter's Notes to Restatement, Second, Torts § 402A, suggests that a product is unreasonably dangerous if it is "dangerous to an extent beyond that which would be contemplated by the ordinary consumer who purchases it, with the ordinary knowledge common to the community as to its characteristics." For the purposes of applying the consumer expectations test to toxic tort claims, "knowledge common to the community" may connote either knowledge held generally by contemporary adults, or it may mean knowledge of the hazards or propensities of a product that professional users might typically have. Evaluation of the reasonable expectations of adult members of the general consuming public would be appropriate for toxic harms arising from household or home gardening goods, ranging from appliance radiation to lawn fungicides. Where, on the other hand, a potentially hazardous product is used principally or exclusively by professional users, knowledge of the risks and characteristics of a toxic product would be that presumably held by others in that professional, industrial, or other context.

The effect of targeting the "consumer expectations" of a professional community is that, for example, a manufacturer might not be found to have failed to provide adequate warnings or instructions as to the hazards of skin contact with roofing tar if the tar involved in a workplace injury was sold only in large containers marketed to professional roofing workers who, it could be fairly presumed, know of the product's risks and the means of avoiding those risks. While courts reiterate that plaintiff's strict liability count "focus[es] . . . on the consumer's expectations" rather than on the conduct of the manufacturer,[8] there is nevertheless general agreement that application of the consumer expectation test in strict liability involves considerations of risk and foreseeability that are similar to those used in negligence analysis.[9]

Comment i to § 402A suggests that some arguably deleterious products, such as tobacco or distilled spirits, are not dangerous to an extent beyond which the purchaser of ordinary knowledge and awareness would expect and are, therefore, not "unreasonably dangerous".[10] Other courts have not interpreted comment i's reference to tobacco products to preclude *any* action against manufacturers of smoking products unless plaintiff can show product adulteration. Under this view, an unreasonably dangerous defect, i.e., a condition falling below the expectation of the ordinary consumer, may arise in a tobacco product by virtue of its "design, packaging and manufacturing variations".[11]

7.3.3 Risk-Utility Test

The "risk-utility" test for measuring risks posed by a toxic product requires balancing the risks inherent in a product, as formulated, against its utility and cost. Factors selected from a typical statement of the risk-utility test include: (1) the utility of the product to the public; (2) the likelihood the product will cause injury; (3) the availability and commercial feasibility of a safer alternative product; and (4) plaintiff's ability to avoid any hazard by careful use of the product.[12]

A conventional description of the risk-utility evaluation requires plaintiff to demonstrate that the magnitude of the risk involved in marketing the product in its challenged formulation outweighed the manufacturing and social costs of changing the product. For example, in *Dine v. Western Exterminating Co.*,[13] an action against manufacturers of chlordane and heptachlor, chemicals used as termiticides, the court adopted a risk utility evaluation providing that "a manufacturer is strictly liable for damage caused by his product if there was a feasible way to design a safer product and an ordinary consumer would conclude that the manufacturer ought to have used that alternative design. To establish this, a plaintiff must show that the magnitude of the danger from the product outweighed the costs of avoiding the danger — including, for example, any new dangers created and any reduction in the benefits of the product caused by the new design."[14]

7.3.4 State of the Art

Generally, a manufacturer is responsible to produce a product that is reasonably safe for its intended use and for reasonably foreseeable misuses. The congruent proposition is that the relative safety with which a product can be used may be a function of the state of scientific and technical knowledge in existence at the time the product's design or formulation was adopted, and at the time the product was produced and offered for sale. Accordingly, in the strict liability claim of most toxic tort suits, the manufacturer or seller has been permitted to produce evidence that the design of a product at the time of its original introduction into commerce was consistent with the "state of the art."[15] The "state of the art" comprises the then-existing state of scientific and industry knowledge as to reasonably safe production and use of the product, consistent with risks known or foreseeable at the time.

7.4 DEFENSES

7.4.1 Generally

The toxic tort plaintiff bringing a claim in strict products liability must show that the product was being used in a normal or a reasonably foreseeable manner. Where the user's employment of the product is starkly incongruous with ordinary use, the manufacturer or seller will raise the defenses of (1) product misuse or (2) assumption of the risk. In most jurisdictions, the defenses of product misuse or assumption of the risk no longer operate to bar plaintiff's recovery altogether, but rather serve as evidence that may reduce plaintiff's damages pursuant to a comparative fault evaluation. Of course, where plaintiff's conduct is so incautious or reckless that it can no longer be said that the condition of defendant's product was the proximate cause of plaintiff's harm, product misuse or assumption of the risk may serve as a complete defense to plaintiff's claim. A fuller discussion of the toxic tort defendant's responses to plaintiff's claims is found in Chapter 14.

7.4.2 Contributory Negligence

In a strict products liability suit, contributory negligence of the plaintiff "is not a defense when such negligence consists merely in a failure to discover the defect in the product, or to not guard against the possibility of its existence."[16] The rationale for not permitting contributory negligence as a defense in a strict liability suit is that as the conduct of the manufacturer or seller is not at issue in a strict tort suit, it would be inharmonious to permit plaintiff's absence of due care to preclude recovery.

7.4.3 Assumption of the Risk

Assumption of the risk is a more aggravated form of plaintiff misconduct and may constitute a defense to plaintiff's strict liability claim. Where plaintiff's conduct goes beyond contributory negligence, and instead consists of "voluntarily and unreasonably proceeding to encounter a known danger", this conduct, known as assumption of the risk, may bar plaintiff's claim.[17] Consistent therewith, it has been held that a cigarette smoker who continues to smoke notwithstanding widespread knowledge of smoking risks will be barred from recovery for smoking-related disease.[18]

7.4.4 Product Misuse

In strict tort products liability, as in negligence, the manufacturer may be liable for injuries arising from an unreasonably dangerous condition of the product that causes injury when the product is used in the manner intended or, even if not intended, a manner that is reasonably foreseeable. Even where plaintiff's misuse of a toxic product is conceded, it will not bar tort recovery where, for example, the manufacturer's warnings or instructions failed to provide adequate guidance as to the presence of the risk or the means of avoiding it. For example, in one suit plaintiffs injured after mixing a cleaning solution in a stronger proportion that recommended by the product's label were not barred from recovery where the product's label stated that the cleaning solution was harmless,[19] nor were plaintiffs injured by carbon monoxide after using charcoal cooking briquettes as an indoor fuel precluded from recovering from defendant where the product failed to alert them of the risks of that use.[20]

NOTES

1. 377 P.2d 897 (1962).
2. Restatement, Second, Torts § 402A.
3. Restatement, Second, Torts § 402 ("Special Liability of Seller of Product for Physical Harm to User or Consumer") provides:

   (1) One who sells any product in a defective condition unreasonably dangerous to the user or consumer or to his property is subject to liability for physical harm thereby caused to the ultimate user or consumer or to his property, if
   (a) the seller is engaged in the business of selling such a product, and
   (b) it is expected to and does reach the user or consumer without substantial change in the condition in which it is sold.
   (2) The rule stated in Subsection (1) applies although
   (a) the seller has exercised all possible care in the preparation and sale of his product, and
   (b) the user or consumer has not brought the product from or entered into any contractual relation with the seller.

4. E.g., *Gilboy v. American Tobacco Co.*, ____ So. 2d ____, Prod. Liab. Rep. (CCH) ¶ 12,854 (La. 1991).
5. See, e.g., *Leng v. Celotex Corp.*, 196 Ill. App. 3d 647, 554 N.E.2d 468, Prod. Liab. Rep. (CCH) ¶ 12,472 (Ill. App. 1990), where the court, in an action brought by the estate of a worker who died of mesothelioma following extensive exposure to asbestos, states the general rule in this way: "In a products liability action, plaintiff must establish that the injury complained of resulted from a condition in the product, that the condition was unreasonably dangerous, and that the condition existed when the product left the manufacturer's control." *Id.*, 196 Ill. App. 3d at 649, 554 N.E.2d at 469.
6. E.g., *Zumo v. R.T. Vanderbilt Co., Inc.*, 527 So. 2d 1074 (La. App. 1988) (declining to grant summary judgment to manufacturer where plaintiff worker raised material issue of fact concerning the unreasonably dangerous formulation of the challenged chemical products).
7. *Enright v. Eli Lilly & Co.*, 77 N.Y.2d 377, 586 N.Y.S.2d 550, 570 N.E.2d 198, Prod. Liab. Rep. (CCH) ¶ 12,734 at 38,266 (N.Y. 1991), *cert. denied*, ____ U.S. ____ (1991).
8. *Crislip v. TCH Liquidating Co.*, 52 Ohio St. 3d 251, 556 N.E.2d 1172 (Ohio 1990) (carbon monoxide poisoning from use of wood burning add-on furnace).

9. *Crislip v. TCH Liquidating Co., supra,* 556 N.E.2d at 1182, citing 1 Madden, Products Liability (2nd ed. 1988) §§ 10.1–10.3; *Phillips v. Kimwood Machine Co.,* 269 Ore. 485, 493, 525 P.2d 1033, 1037 (1974) ("[A] manufacturer who would be negligent in marketing a given product, considering its risks, would necessarily be marketing a product which fell beneath the reasonable expectations of consumers who purchased it.").
10. "Good whiskey is not unreasonably dangerous merely because it will make some people drunk, and is especially dangerous to alcoholics; but bad whiskey, containing a dangerous amount of fusel oil, is unreasonably dangerous. Good tobacco is not unreasonably dangerous because the effects of smoking may be harmful; but tobacco containing something like marijuana may be unreasonably dangerous." Restatement, Second, Torts § 402A, comment i. A more contemporary example of the latter reference to adulterated tobacco would be recent litigation in which it is claimed that years ago one or more cigarette manufacturers used asbestos as a constituent element of a cigarette filter.
11. *Rogers v. R.J. Reynolds Tobacco Co.,* 557 N.E.2d 1045, 1053 n.8, Prod. Liab. Rep. (CCH) ¶ 12,572 at 37,531 n.8 (Ind. App. 1990) ("Cigarettes, though they contain tobacco, are manufactured products. As such, they are subject to design, packaging, and manufacturing variations which may indeed render them defective though the tobacco they contain is unadulterated. Failing to warn the consumer of certain properties, such as nicotine addiction, in conjunction with its harmful qualities, render the product unreasonably dangerous, i.e., in a condition the ordinary consumer would not contemplate. Also, a design defect which renders a product more addictive than it could be or addictive when it need not be at all may render the cigarette unreasonably dangerous in conjunction with its harmful qualities. Defects such as the foregoing satisfy the element that 'something be wrong' with the cigarette.").
12. See, e.g., *Dine v. Western Exterminating Co.,* Prod. Liab. Rep. (CCH) ¶ 11,714 (D.D.C. 1988) (not reported in F. Supp.). *Cf., Cepeda v. Cumberland Engineering Co.,* 76 N.J. 152, 386 A.2d 816, 825–829 (N.J. 1978) ("(1) the usefulness and desirability of the product — its utility to the user and to the public as a whole; (2) The safety aspects of the product — the likelihood that it will cause injury, and the probable seriousness of the injury; (3) The availability of a substitute product which would meet the same need and not be as unsafe; (4) The manufacturer's ability to eliminate the unsafe character of the product without impairing its usefulness or making it too expensive to maintain its utility; (5) The user's ability to avoid the danger by the exercise of care in the use of the product; and (6) The user's awareness of the dangers inherent in the product and their avoidability, because of the general public knowledge of the obvious condition of the product, or the existence of suitable warnings or instructions.").
13. Prod. Liab. Rep. (CCH) ¶ 11,714 at 33,406 (D.D.C. 1988).
14. *Id.* See also *Zumo v. R.T. Vanderbilt Co., Inc.,* 527 So.2d 1074 (La. App. 1988) (worker raised material issues of fact as to whether dangers inherent in chemical outweighed its utility).
15. E.g., *Anderson v. Owens-Corning Fiberglas Corp.,* 810 P.2d 549 (Cal. 1991).
16. Restatement, Second, Torts § 402A, comment n.
17. Restatement, Second, Torts § 402A, comment n.

18. See Generally Note, "Plaintiff's Conduct as a Defense to Claims Against Cigarette Manufacturers," 99 Harv. L. Rev. 809 (1986).
19. *Cully v. Fuller Brush Co.*, 68 Wash. 2d 675, 415 P.2d 7 (1966).
20. *Alfieri v. Cabot Corp.*, 17 A.D.2d 455, 235 N.Y.S.2d 753 (1963), affirmed 13 N.Y.2d 1027, 245 N.Y.S.2d 600, 195 N.E.2d 310 (1963).

CHAPTER 8

Liability for Abnormally Dangerous Activities

CONTENTS

Liability for Abnormally Dangerous Activities

8.1 GENERALLY

Some products create such grave risks when accidentally released from the control of the manufacturer, transporter, or user, that the responsible party may be liable even where they exercised scrupulous care.[1] This doctrine was once described as ''liability for ultrahazardous activities,'' although the modern definition uses the term ''abnormally dangerous''.

As is readily apparent, it is to plaintiff's substantial advantage to succeed in having defendant's conduct characterized as ''abnormally dangerous''. With such designation, plaintiff can prove liability without having to prove defendant's culpable conduct, i.e., defendant will be strictly liable for plaintiff's damages without the need for plaintiff to prove that defendant acted intentionally, recklessly, or negligently. As will be equally clear from the discussion that follows, however, only a small cluster of potentially hazardous activities have been designated ''abnormally dangerous''.

To show that defendant's activity is ''abnormally dangerous'', plaintiff's burden is substantially greater than a simple showing that defendant's pursuits carry with them the possibility of great harm. Rather, plaintiff must show (1) the risk of an abnormally great harm should defendant's safety efforts fail; (2) the virtual impossibility of defendant's elimination of the risk of harm even with the utmost care; and (3) a resultant harm to plaintiff, or plaintiff's property, caused by the very hazards the risk of which led to describing defendant's conduct as ''abnormally dangerous'' in the first instance.

8.2 RESTATEMENT, SECOND, TORTS §§ 519 AND 520

Restatement, Second, Torts § 519 states that ''[o]ne who carries on an abnormally dangerous activity is subject to liability for harm to the person, land, or chattels of another resulting from the activity, although he has exercised the utmost care to prevent the harm.'' The section concludes with the proviso that ''[t]his strict liability is limited to the kind of harm, the possibility of which makes the activity abnormally dangerous''.

Section 520 suggests the factors to be considered in determining whether a particular activity is abnormally dangerous. The factors include ''(a) the existence of a high degree of risk of some harm to the person, land, or chattels of others; (b) likelihood that the harm that results from it will be great; (c) inability to eliminate the risk by the exercise of reasonable care; (d) extent to which the activity is not a matter of common usage; (e) inappropriateness of the activity to the place where it is carried on; and (f) extent to which its value is outweighed by its dangerous attributes.''

8.2.1 Danger Unavoidable Even with the Exercise of Due Care

In a toxic tort suit alleging an abnormally dangerous activity, a plaintiff must successfully demonstrate more than the high risk posed by defendant's activity, its incongruity with the surrounding commercial or industrial environment, and its capacity for great harm. Liability for an abnormally dangerous activity will not lie unless plaintiff makes the additional showing that the risk involved cannot be eliminated through defendant's exercise of reasonable care.

Illustrative is the decision in *Edwards v. Post Transportation Co.*,[2] the suit by an employee against the delivery company that pumped sulfuric acid into the wrong storage tank at the manufacturer's waste-treatment facility. The introduction of sulfuric acid into a tank containing a residue of sodium bisulfate and water created toxic gases that overcame and severely injured plaintiff. Turning to plaintiff's allegation of liability for abnormally dangerous activities, the California appellate court agreed that plaintiff's proof satisfied several of the criteria listed by the Restatement.[3] It held, nonetheless, that plaintiff's claim under Restatement, Second, Torts §§ 519 and 520 must fail because the evidence showed that the risk of the misadventure resulting in plaintiff's harm could be ''eliminated through the exercise of reasonable care.'' In so concluding, the court noted the testimony of one of plaintiff's expert witnesses, a civil engineer, who testified that ''[i]f sulfuric acid is handled in a proper fashion, it is no danger,'' leading the court to the ''logical conclusion'' that the hazard ''can be eliminated through care.''[4]

8.2.2 Requirement of an Activity

Claimants proceeding against product manufacturers for claims arising from the manufacture and marketing of products claimed to be abnormally dangerous have been denied recovery under this doctrine for want of demonstrating the ''activity'' requirement of plaintiff's prima facie case. For example, in *Heinrich v. Goodyear Tire and Rubber Co.*,[5] a tire company employee brought an action against a chemical supplier for injuries sustained due to exposure to the chemical. The federal trial court held that the doctrine of liability for abnormally dangerous activities would not lie, stating that ''it is apparent that the party to be held liable under such a theory must have at least the right or duty to control, if not actual control over, the activity causing the harm.''[6] A similar result was reached in a suit brought by a plant employee who was sprayed by liquified anhydrous ammonia. In that suit, plaintiff proceeded against the plant's former owner, the plant's designer and installer, and the manufacturer of the valvė that had misfunctioned. Recovery was denied on the grounds that the named defendants were not operating the plant at the time of the accident.[7]

Likewise, courts considering the question have held that the manufacturer and marketing of cigarettes is not an abnormally dangerous activity within the meaning of the Restatement, Second, Torts §§ 519 and 520. For example, in *Gilboy v. American Tobacco Co.*,[8] a suit arising from use of tobacco products, the court stated that "ultrahazardous activities contain risks that are reasonable but still dangerous enough that a party undertaking these activities will assume the consequences for injury to others," adding that smoking is not related "in any way to an ultrahazardous activity" on the part of defendants.

8.2.3 Application of Doctrine

Various courts have evaluated plaintiff's claims under the doctrine of abnormally dangerous activities, with results either recognizing or denying relief for activites ranging from the transportation to the inadvertant emission of toxins.

In one suit involving emission of sodium hydroxide, the court concluded that even where plaintiff succeeds in proving that defendant's industrial activity resulted in harmful emissions, an action will not lie under Restatement, Second, Torts § 519-520 without proof that defendant's activity "posed a great danger of invasion of the land of others." A New York appellate court so held in *DeFoe Corp. v. Semi-Alloys, Inc.*,[9] a suit brought against the manufacturer of electronic components by an adjoining business that alleged that the malfunction of defendant's "scrubber" caused "the emission of considerable quantities of sodium hydroxide." Affirming the trial court's dismissal of plaintiff's abnormally dangerous activities count, the appellate court held that even though sodium hydroxide is an active caustic chemical, it posed an insufficiently great risk to nearby landowners, unlike other activities including, inter alia, "using atomic energy". The court added that "plaintiff made no showing that the defendant would be unable to eliminate the risk by the exercise of reasonable care, . . . or that the activity the defendant was involved in was inappropriate in the place it was carried on[.]"[10]

Some judicial commentary includes the risk of groundwater contamination from stored fuel as warranting strict tort liability for abnormally dangerous activities. In *Siegler v. Kuhlman,*[11] the Washington Supreme Court applied strict liability to a defendant responsible for gasoline spillage on a highway. While emphasizing the risk of gasoline "volatility, inflammability and explosiveness",[12] the court also noted that "stored in commercial quantities, gasoline has been recognized to be a substance of such dangerous characteristics that it invites a rule of strict liability — even where the hazard is contamination to underground water supply and not its more dangerous properties such as its explosiveness and flammability."[13]

Abnormally dangerous activities liability may be applied to transporters of toxic substances. In one suit, a federal trial court found that plaintiff's complaint stated sufficient allegations to permit it to prove Restatement, Second, Torts §§ 519 and 520 liability against the manufacturer/shipper of the hazardous and toxic substance acrylonitrile which leaked from defendant's freight car, causing damage to property, equipment, and the adjacent water supply.[14]

In most suits, on the other hand, accidents arising from "the transmission on land of natural gas and petroleum products by pipeline, train, or motor vehicle, [as] a common activity in a highly industrialized society such as our own[,]" have been held unsuited to application of strict liability for abnormally dangerous activities.[15]

It has also been held that a plaintiff may state a claim in strict liability for abnormally dangerous activities arising from the release of gamma radiation and radon from radium processor waste tailings;[16] contamination of adjoining property by a landowner's ground-based spraying of herbicides[17] and the explosion of a waste-oil tanker truck in the course of a welding operation.[18]

8.2.4 Affirmative Defenses

Defendant in an abnormally dangerous activities claim may avoid liability by showing that the harm was caused by the nonforeseeable intervening act of a third party. For example, in *Davis Enterprises v. Gouza,*[19] a suit brought by residents of a housing development against a petroleum pipeline company and a cable company engaged in excavation work for personal injury and property damage arising from the spillage of 50,000 gallons of unleaded gasoline, stated: "the leak was allegedly caused by the intervening negligence of Tri-State Communications, Inc., and the obligation of Sun Pipe Line Company does not rise to that of an insurer."[20]

NOTES

1. "An activity is ultrahazardous if it (a) necessarily involves a risk of serious harm to the person, land or chattels of another which cannot be eliminated by the exercise of the utmost care, and (b) is not a matter of common usage . . . " *Luthringer v. Moore,* 190 P.2d 1, 7 (1948) (fumigation).
2. 228 Cal. App. 3d 980, 279 Cal. Rptr. 231 (1991) (review denied).
3. "We are satisfied, for instance, that sulfuric acid is 'not a matter of common usage.' . . . . Sulfuric acid, though unquestionably a useful and beneficial chemical, is not so necessary that we should insulate its users from strict liability[.] . . . We conclude that the plaintiff, by a preponderance of the evidence satisfied Restatement factor (a) — that the use of sulfuric acid involves a 'high degree of risk,' and factor (b) — that if it comes into contact with

humans the resulting harm is likely to be great." *Id.*, 228 Cal. App. 3d at 986, 279 Cal. Rptr. at 233–234.

4. "Where the activity is dangerous only if insufficient care is utilized, ordinary rules of fault are sufficient for allocation of the risk. There is no need for liability without proof of fault, because definitionally if there is some danger it will have resulted from negligence and be compensable." *Id.*, 228 Cal. App. 3d 980, 279 Cal. Rptr. 231, 234-235.
5. 532 F. Supp. 1348 (D. Md. 1982).
6. *Id.*, 532 F. Supp. at 1357.
7. *Cropper v. Rego Distribution Center, Inc.*, 542 F. Supp. 1142, 1149 (D. Del. 1982).
8. 572 So. 2d 287 (La. App. 1990), reversed on other grounds.
9. 549 N.Y.S.2d 133 (N.Y. App. Div. 1989).
10. *Id.*, 549 N.Y.S.2d at 135.
11. 81 Wash. 2d 448, 502 P.2d 1181 (Wash. 1972), *cert. denied* 411 U.S. 983 (1973).
12. *Id.*, 502 P.2d at 1184–1185.
13. Id., 502 P.2d at 1186. See also *Yommer v. McKensie*, 255 Md. 220, 257 A.2d 138 (1969). In that suit, evaluating a claim brought by a family in a semi-rural community against the owner of gasoline storage tanks, claiming contamination of the household's drinking water, the court stated: "The thing that stands out from the cases is that the important thing about the activity is not that it is extremely dangerous in itself, but that it is abnormally so in relation to its surroundings. . . . The same is true of the storage of gasoline, or other inflammable liquids, in large quantities. In a populated area, this is a matter of strict liability [citations omitted]. But in an isolated area it is not [citations omitted]." *Id.*, 257 A.2d at 141.
14. *Indiana Harbor Belt Railway Co. v. American Cyanamid Co.*, 517 F. Supp. 314 (N.D. Ill. 1981). See also *China Union Lines Ltd. v. A.O. Anderson & Co.*, 364 F.2d 769 (5th Cir. 1966), in which the court suggested the possibility of abnormally dangerous activities liability for a shipper of acrylonitrile, and remanded the action for trial court determination of "what Cyanamid actually knew or ought to have known about its own product, and knowing that, what in prudence considering these extremely high hazards could and ought to be done to minimize, if not eliminate, the devastating consequences of a foreseeable causality." *Id.*, 364 F.2d at 797.
15. *Davis Enterprises v. Gouza*, 576 A.2d 999, 587 (Pa. Super. 1990) (rupture, during excavation operations by third party, of pipeline carrying unleaded gasoline).
16. *T & E Industries, Inc. v. Safety Light Corp.*, 227 N.J. Super. 228, 546 A.2d 570 (App. Div. 1988), certification granted 117 N.J. 119, 564 A.2d 848 (1989).
17. *Speer & Sons Nursery, Inc. v. Duyck*, 759 P.2d 1133 (1988), review denied 765 P.2d 814 (Or. App. 1988).

18. *Garcia v. Estate of Norton*, 183 Cal. App. 3d 413, 228 Cal. Rptr. 108 (1986).
19. 576 A.2d 999 (Pa. Super. 1990).
20. *Id.*, 576 A.2d 999, 1003–1004.

CHAPTER 9

The Duty to Warn

CONTENTS

The Duty to Warn

9.1 GENERALLY

The rule is universal that "a product may be defective because of the absence of an adequate warning of the dangers inherent in its use."[1] Even "a well-designed product may be deemed defective without an adequate warning of dangers inherent in its normal or reasonably foreseeable use which a user otherwise might not fully appreciate."[2] Conversely, evidence that plaintiff knowingly disregarded adequate warnings provided by a manufacturer will ordinarily preclude recovery for failure to warn.[3]

As a basic proposition, the manufacturer is only responsible for providing warnings or instructions regarding risks, products uses, and exposures that are reasonably foreseeable. Accordingly, a claimant whose exposure to the toxin was purely fortuitous may be denied recovery for failure to warn.[4] Nonetheless, even a perceptible risk of injury as slight as one in a million may trigger a duty to warn where the injury, should it occur, is significant.[5]

Several factors are pertinent to determining whether the lack of a warning, or an inadequate warning, renders a product unreasonably dangerous and defective. Significant weight is given to (1) the normal expectations of the consumer as to how a product will perform; (2) the nature and magnitude of the danger to which the user is exposed; and (3) the likelihood of injury.

In most circumstances warning adequacy is a matter of fact reserved for determination by the jury. Where, however, the warning is "accurate, clear and unambiguous", the adequacy of a warning "can become a question of law[,]"[6] permitting the court to withdraw the issue of warning adequacy from the jury.[7]

9.2 THE DUTY TO WARN UNDER DIFFERENT CLAIMS

9.2.1 Negligence

The product manufacturer's duty is not to ensure that no person will be injured by use of or exposure to its products. Rather, the duty is to manufacture products that are reasonably safe for their intended use or reasonably foreseeable misuse. Under negligence law, a product manufacturer has a duty, at a minimum, to inform foreseeable users or consumers of product hazards that may not be readily apparent.

Where the manufacturer fails to do so, and where that failure to warn or provide adequate instructions proximately causes plaintiff's personal injury or property damage, the manufacturer may be liable in tort for damages to compensate plaintiff for the loss. The rationale for finding negligence liability in such circumstances is that a product manufactured and sold without warnings or instructions is unreasonably dangerous and that sale of an unreasonably dangerous product is a breach of the manufacturer's duty of due care.

9.2.2 Warranty

While toxic tort claims may involve warranty claims sounding in express warranty or the implied warranty of fitness for a particular purpose,[8] a toxic tort claim including a claim of failure to provide adequate warnings will typically target breach of the implied warranty of merchantability. That warranty, it is recalled from the earlier discussion, accompanies the sale of a product irrespective of any representations the seller may have made. Where not expressly and conspicuously disclaimed,[9] a seller warrants that the product is fit for its ordinary purpose.

In toxic tort claims involving a claimed failure to provide adequate warnings or instructions, it is generally agreed that a product that creates an unreasonable risk of personal injury or property damage when sold without adequate warnings is unreasonably dangerous. The parallel assessment, accepted by most courts, is that a product sold in such an unreasonably dangerous condition is not fit for its ordinary purpose. Therefore, a manufacturer's sale of a product without adequate warnings breaches the implied warranty of merchantability.

9.2.3 Strict Liability in Tort

In a majority of jurisdictions, the defendant seller is liable in strict tort products liability where it poses an unreasonable risk of injury due to inadequate warnings or instructions. Restatement, Second, Torts § 402A comment j states the seller's obligation to inform the user of hazards of which the seller either knew or should have known at the time of initial sale.[10] Even a product that is unerringly formulated and well suited for its intended use may be deemed unreasonably dangerous without a warning that by its size, location, and effectiveness of language or symbol impresses upon the reasonably prudent user the nature and extent of the hazard involved. A hazard can be great either in terms of the severity of the injury that might occur or in terms of the number of persons who might be adversely affected by the product or substance. The greater the risk, the more prominent, effective, and informative warnings and instructions must be.

As is true in analysis of defendant's duty to warn in warranty and in negligence, the strict liability warning obligation turns on the reasonable foreseeability of a substantial risk.[11] The seller is only under a duty to warn of those dangers that are reasonably foreseeable, prompting the inquiry into "whether it was reasonably foreseeable to the manufacturer that the product would be unreasonably dangerous if distributed without a warning on the label and, if so, whether the manufacturer supplied the warning that a reasonably prudent manufacturer would have supplied.''[12]

9.3 EFFECT OF OBVIOUSNESS OF THE RISK

Most jurisdictions recognize some variation of the so-called ''open and obvious'' defense to a claim that the manufacturer breached its duty to warn. This defense posits that there exists no duty to warn of obviously hazardous conditions. It bears noting, however, that even where a product's hazards may be obvious, if the means of lessening the hazard or avoiding it are not equally obvious, the seller may still have a duty to warn or offer instructions to the user or consumer.

Toxic tort claims concerning injuries claimed to be associated with the use of tobacco products have given courts ample opportunity to recognize the manufacturer's defense that a claimed risk was open and obvious. A conclusion that the risk of using tobacco products is obvious has supported many holdings that cigarette manufacturers have no common law duty to provide consumer warnings in addition to those conveyed pursuant to the the Cigarette Labeling Act.[13]

It is worth noting that in other cigarette products liability actions courts have held the open and obvious rule inapplicable, concluding that there is nothing ''directly observable or deducible from observation'' that would lead a cigarette user ''to conclude that it is either carcinogenic or addictive.''[14]

Of relatively recent origin are suits brought against manufacturer's of alcoholic beverages alleging that the manufacturer failed to adequately warn about the hazardous propensities of use of its products. Such suits generally claim that either the seller should have warned about the risks of personal injury to the consumer or others associated with drinking, or, more specifically, should have warned alcoholics of the dangers of using defendant's products.[15] The Texas Supreme Court disallowed such a claim in *Joseph E. Seagram & Sons, Inc. v. McGuire,* stating that ''because the danger of developing the disease of alcoholism from prolonged and excessive consumption of alcoholic beverages is and has been generally known and recognized, it is within the ordinary knowledge common to the community.''[16]

9.4 CAUSATION

In *Van Buskirk v. Carey Canadian Mines, Ltd.,*[17] involving claims of asbestos-related injury, the federal appeals court explained that ''plaintiff must establish causation by showing that had defendants issued a proper warning, plaintiffs would have altered their behavior to avoid the injury.'' Defendants, in turn, may defeat plaintiff's claim of causation ''by discrediting plaintiffs' claims that they would have acted to avoid injury, or by pointing to a third party as the sole proximate cause.''[18]

Two presumptions are recognized in many toxic tort claims of failure to provide adequate warnings. Restatement, Second, Torts § 402A, comment j, states that ''[w]here a warning is given, the seller may reasonably presume that it will be read and heeded[.]'' The converse of this rule, recognized in many states, is that the plaintiff may avail himself of the presumption that had an adequate warning been given, plaintiff would have conformed his or her behavior so as to avoid the risk.[19]

9.5 PROFESSIONAL, SOPHISTICATED, OR KNOWLEDGEABLE USERS

In the context of a sale of a toxic product to an industrial user, employees of the purchaser often bring claims against the manufacturer for failing to warn adequately about the risks of product exposure. In settings where the purchaser knows of the salient toxic risks, decisional law has carved a limited exception to the general rule that a manufacturer must provide warnings and instructions directly to the users (in these cases, employees of the purchaser). A manufacturer seeking to avail itself of the professional or sophisticated user defense must show that (1) it apprised the buyer-employers of the risks posed by use of or exposure to the product and (2) it ascertained that the buyer-employer could be reasonably relied upon to provide adequate warnings to the employees.[20]

For example, in *Smith v. Walter C. Best, Inc.*,[21] a federal trial court, applying Ohio law, applied a ''knowledgeable purchaser'' defense to defeat a claim by a foundry worker against suppliers of silica sand to the worker's employer and found that defendant suppliers were entitled to rely upon the foundry to warn its workers of the risks of prolonged inhalation of silica dust. A federal trial court applied this doctrine in *Byrd v. Brush Wellman, Inc.*,[22] a suit brought against a manufacturer of beryllium oxide by an employee of the product's purchaser. The employee had contracted berylliosis resulting from his exposure to the product and claimed that the manufacturer had breached its duty to warn of exposure limitations and means of avoiding the risk. Citing authority holding that a product supplier has no duty ''to warn employees of [a] knowledgeable industrial purchaser as to product related hazards'',[23] the court found that the manufacturer's adequate warnings to the industrial purchaser, coupled with the purchaser's ample knowledge of beryllium hazards, required the conclusion that the manufacturer's conduct did not proximately cause the employee's injury.[24] A like result was reached in the asbestos case *GAF Corp. v. United States*,[25] which held that the federal government should not be liable for failure to warn a government contractor of the hazards of handling insulation products. The court reasoned that the claimed ''superior knowledge'' of the government as to the risks did not impose upon it a duty to inform an experienced producer of the risks posed by its own products.

Bulk suppliers may be exempted from a duty to warn all persons who might ultimately come into contact with a toxin where the supplier has provided sufficient safety warnings to immediate distributees.[26]

9.6 ADEQUACY OF WARNING

A toxic tort claimant raises a triable issue of warning adequacy where "[r]easonable people might disagree about whether any or all of the warnings . . . were unambiguous and conspicuous, and whether they provided sufficient information on the extent of risks and symptoms of overexposure to defendant's products".[27] Often a manufacturer's warnings at the time of initial sale were arguably adequate to apprise users of foreseeable risks, but at a later point the manufacturer becomes aware of risks in use or application of the product not previously known. In such circumstances, a manufacturer may be held to a continuing informational duty, sometimes called "a post sale duty to warn".[28] To be contrasted, a manufacturer who has effected product improvements in successive product designs need not advise purchasers of the improved product of the potential defects in the earlier version.[29]

9.7 DEFENSES

Under negligence, warranty, and strict tort liability, the informational duties of the manufacturer of toxic products are to provide adequate warnings as to reasonably foreseeable risks. Accordingly, a manufacturer has no duty to warn of toxic hazards that were unknown at the time the product was initially distributed into commerce.

Consistent with this general rule, some defendants in asbestos cases still argue successfully that industry and scientific knowledge from the late 1930s to the 1960s permit the conclusion that they neither knew nor should have known of the risks of asbestos exposure during these earlier time periods.[30]

Defendants in asbestos-related injury suits have had difficulty in proving that claimant knowledge of risks or other unhealthful habits, such as cigarette smoking, should absolve defendants of liability. In one suit, an asbestos manufacturer failed in its efforts to demonstrate claimant assumption of risk in a suit brought by an asbestos worker who continued to smoke cigarettes after the appearance of mandatory warning labels on tobacco products.[31] In another, *Skonberg v. Owens-Corning Fiberglas Corp.*,[32] an appeals court held that the evidence at trial supported the jury's finding that the manufacturer's failure to warn of asbestos risks proximately caused an insulation worker's lung cancer, even in light of evidence that the worker, a smoker, (1) ignored cigarette package warnings; (2) was advised, through union publications, of asbestos risks; and (3) had discussed, with his wife, the harmful effects of asbestos exposure.

NOTES

1. *Jackson v. Deft, Inc.*, 223 Cal. App. 3d 1305, Prod. Liab. Rep. (CCH) ¶ 12,728 at 38,246 (1990)(paint fumes), citing *Brown v. Superior Court*, 44 Cal. 3d 1049, 1057, 245 Cal. Rptr. 412, 751 P.2d 470 (1988) (pharmaceuticals).
2. *Rastelli v. Goodyear Tire & Rubber Co.*, 565 N.Y.S.2d 889 (N.Y. App. Div. 1991) (action against several manufacturers of multipiece tire rims).
3. E.g., *Richards v. C. Schmidt Co.*, 54 Ohio App. 3d 123, 1989 WL 225788 (Ohio App. 1989) (foam machine worker's failure to use mask or respirator even though worker understood manufacturer's warnings to mean that it was hazardous to breath chemical vapors).
4. *Rivers v. AT&T Technologies, Inc.*, 554 N.Y.S.2d 401, 1990 WL 48718 (N.Y. Sup. 1990) (file clerk in telephone company's control center denied recovery from bulk supplier of chemical solvent dimethylformamide (DMF)).
5. See *Graham v. Pittsburgh Corning Corp.*, 593 A.2d 567 (Del. Sup. 1990). In that suit, brought against a manufacturer of asbestos products, the court states: "American Law of Products Liability 3d § 32:3 (1987) states the duty to warn exists even 'where the possibility of injury is very small, as little as only one in a million, where the potential injury is great.' *Prosser and Keeton on Torts* (W. Keeton 5th ed. 1984) § 96 and 1 Madden, *Products Liability* (2nd ed. 1988) § 10.1 require warning where the product poses unreasonable risk to the user." 593 A.2d at 569.
6. *Upjohn Co. v. MacMurdo*, 562 So. 2d 680, Prod. Liab. Rep (CCH) ¶ 12,602 at 37673 (Fla. 1990) (contraceptive manufacturer's warnings about the risks of abnormal bleeding), quoting *Felix v. Hoffman-LaRoche, Inc.*, 540 So. 2d 102, 105 (Fla. 1989).
7. See generally *Hall v. Ashland Oil Co.*, 625 F. Supp. 1515 (D. Conn. 1986).
8. Toxic tort claims based in warranty are discussed in Chapter 6.
9. The requisites of warranty disclaimers are discussed in Chapter 6.
10. "In order to prevent the product from being unreasonably dangerous, the seller may be required to give directions or warning, on the container, as to its use." Restatement, Second, Torts § 402A, comment j.
11. E.g., *Rodriguez v. Superior Court of Kern County*, 271 Cal. Rptr. 114, 1990 WL 91300 (Cal. App. 1990) (held: plaintiff minor, through guardian ad litem, stated a strict liability cause of action for failure to warn adequately against aspirin manufacturer, the court noting that certain uses of the product involved a substantial and foreseeable risk of Reyes syndrome).
12. *Anderson v. Klix Chem. Co.*, 256 Or. 199, 472 P.2d 806, 808 (1970).
13. E.g., *Varga v. Brown & Williamson Tobacco Corp.*, _____ F. Supp. _____, Prod. Liab. Rep. (CCH) ¶ 12,041 (W.D. Mich. 1988) (prisoner's claim against

manufacturer of Bugler and Kite tobacco products dismissed, the court stating: "[Plaintiff's] complaints and other pleadings, even read liberally, allege only that the failure to include warnings on Bugler and Kite products rendered the products defective. Since the dangers associated with the use of tobacco were obvious in 1986, plaintiff cannot allege a set of facts establishing a duty to warn on the part of defendant.").

14. *Rogers v. R.J. Reynolds Tobacco Co.*, 557 N.E.2d 1045, Prod. Liab. Rep. (CCH) ¶ 12,572 at 37,530 (Ind. App. 1990).
15. *McGuire v. Joseph E. Seagram & Sons, Inc.*, 790 S.W.2d 842 (Tex. App. 1990) (plaintiff, an alcoholic, stated cause of action in failure to warn and for absence of adequate information to enable customers to safely use defendant's products).
16. 814 S.W.2d 385, (Tex. 1991).
17. 760 F.2d 481, 492–493 (3d Cir. 1985).
18. *Id.*, 760 F.2d at 492–493.
19. *Cf.*, *Batteast v. Wyeth Laboratories, Inc.*, 137 Ill. 2d 175, 1990 WL 91104 (Ill. 1990) (evidence in action arising from drug overdose from administration of aminophylline suppositories raised jury issue as to warning inadequacy and causation; evidence indicated that prescribing physician would have either declined to prescribe the drug or more closely monitored its administration had the manufacturer's package insert contained more detailed warnings and instructions).
20. Restatement, Second, Torts § 388 & comment n.
21. 756 F. Supp. 878 (W.D. Pa. 1991).
22. 753 F. Supp. 1403, Prod. Liab. Rep. (CCH) ¶ 12,641 (E.D. Tenn. 1990).
23. *Id.*, ¶ 12.641 at 37,845, quoting *Goodbar v. Whitehead Bros.*, 591 F. Supp. 552, 566–67 (W.D. Va. 1984).
24. *Id.*, ¶ 12,641 at 37,853 ("Given the state of 3M's knowledge regarding beryllium, Brush-Wellman cannot be liable for the injury suffered by the plaintiff. First, Brush-Wellman clearly provided an adequate warning to 3M regarding the dangers of beryllium, both as to the two forms of berylliosis and the dangers of mist containing beryllium particulates. Second, Brush-Wellman's conduct was clearly not the proximate cause of plaintiff's injury. Lastly, since 3M was so obviously knowledgeable about the dangers associate with beryllium-containing products and was the only party in a position to effectively warn the plaintiff and guard against those dangers, Brush-Wellman had no duty to warn the plaintiff.").
25. 932 F.2d 947 (Fed. Cir. 1991).
26. *Rivers v. AT&T Technologies, Inc.*, 554 N.Y.S.2d 401 (N.Y. Sup. 1990) (bulk supplier of chemical solvent dimethylformamide (DMF) where immediate distributees were responsible intermediaries, aware of the risks of exposure to the solvent, and where the supplier had not control over the use of the substance once it was altered for ultimate use).
27. *Jackson v. Deft, Inc.*, 223 Cal. App. 3d 1305, Prod. Liab. Rep. 12,728 at 38,246 (1990).
28. E.g., *Rastelli v. Goodyear Tire & Rubber Co.*, 565 N.Y.S.2d 889 (App. Div. 1991) (manufacturers of multipiece tire rims).

29. *Sill v. Shiley, Inc.*, 735 F. Supp. 337 (W.D. Mo. 1990) (no liability in failure to warn and claimed damages for emotional distress for manufacturer of improved heart valve who did not inform purchaser of flaws causing risk of fracture in previous versions of the valve).
30. *In re Joint Eastern and Southern District Asbestos Litigation*, 762 F. Supp. 519 (E.D.N.Y. 1991) (evidence at trial did not compel conclusion that manufacturers knew or should have known of the hazard during this earlier time period).
31. *Raney v. Owens-Illinois, Inc.*, 897 F.2d 94 (2d Cir. 1990).
32. 215 Ill. App. 3d 735, 576 N.E.2d 28 (Ill. App. 1991).

CHAPTER 10

Causation

CONTENTS

Causation

10.1 GENERALLY

The toxic tort claimant must establish a reasonable connection between the claimed injury or disease and defendant's product or process.[1] In an action alleging injurious exposure at the workplace, plaintiff must show that defendant's product was present at the workplace, that plaintiff worked in proximity to defendant's product, and that plaintiff suffered potentially injurious exposure to defendant's product.[2] Lastly, plaintiff must show by a preponderance of the evidence that exposure to defendant's product was a substantial contributing factor and a direct cause of plaintiff's injury.[3] Accordingly, to show a proximate causal connection between defendant's product and plaintiff's injury or disease, plaintiff must prove more than that defendant's product could conceivably have caused plaintiff's disease.[4]

There are significant recent developments in medical-legal proof of toxic tort causation issues. Of the greatest interest are causation issues raised in claims where it is disputed that defendant's product or process caused, or even could cause, plaintiff's injury.

10.2 BURDEN OF PROOF AND NECESSITY OF EXPERT MEDICAL OPINION

10.2.1 Generally

Where plaintiff's claimed injury is one of medical harm or disease, plaintiff's proof that defendant's product was the proximate cause of the alleged injuries must be by expert testimony.[5] Such testimony "must be stated in terms of reasonable probability."[6] Particular attention is devoted to the basis for an expert's testimony on the question of toxic tort causation,[7] using the logic that in many actions evidence of causation is largely, or even exclusively, the province of expert opinion.[8] It follows that unimpeached expert testimony that plaintiff's exposure to defendant's product could not have caused plaintiff's illness or disease may preclude the claimant from raising a triable issue of proximate cause.[9]

As in claims alleging physical injury, plaintiff's claim for emotional distress due to increased risk will fail absent evidence of actual exposure to the agent claimed to create the risk of injury or disease. Thus, for example, summary judgment was granted against a plaintiff who claimed he was stuck by a needle ineffectively disposed of, although remaining in a device designed for the containment and disposal of used medical syringes. The trial court determined that plaintiff's claim for emotional distress at an increased risk of contracting the AIDS virus could not succeed in the absence of evidence that the needle injuring him had been used in connection with an AIDS patient.[10]

10.2.2 Causation Evidence and Dispositive Motions

Rule 56(c) of the Federal Rules of Civil Procedure provides that summary judgment should be granted only where "there is no genuine issue as to any material fact and . . . the moving party is entitled to judgment as a matter of law."[11] In some suits, expert testimony, through pretrial discovery depositions, has fallen short of establishing a triable issue of fact as to proximate cause. Parties to these suits sought unsuccessfully to avoid summary judgment[12] by interposing additional affidavits in which their experts restated their earlier conclusions in more particular language, or in stronger terms. Most courts have concluded that such tardy initiatives to bolster their expert's testimony are insufficient to avoid summary judgment where "the only issue of fact is to determine which of the two conflicting versions of the . . . testimony is correct."[13]

Where the circumstances of plaintiffs' injury are such as to significantly hamper plaintiff's proof of exposure to defendant's product, some courts require that to avoid summary judgment plaintiff must show only some "significantly probative" evidence in support of the complaint. For example, in *Shields v. Eli Lilly and Co.*,[14] the appellate court overturned a grant of summary judgment to a manufacturer of DES, finding that plaintiff's proof that the prescribing physician sometimes prescribed DES to expectant mothers, and that the description of the pill used made it unlikely that the prescription actually given to the mother was anything but DES.[15] Where there is evidence of a claimant's measurable exposure to a toxin, defendant's production of conflicting evidence will not render the opinions or conclusions of plaintiff's expert inadmissible as not based upon evidence in the trial record.[16]

10.3 PROXIMATE CAUSE

10.3.1 Generally

Proximate cause is often described as "any cause which in a natural and continuous sequence produces the injury and without which the result would not have occurred."[17] Proximate cause is ordinarily an issue for jury determination, although a court may properly grant summary disposition where "plaintiff's evidence does not establish a causal connection, leaving causation to the jury's speculation[.]"[18] Where plaintiff's injury or disease is of a nature that could have been caused by exposure to toxins or conditions not associated with defendant's conduct, plaintiff must produce expert evidence ''isolating other potential causes[.]''[19]

Accordingly, some courts have concluded that the toxic tort claimant's own testimony that she was exposed to defendant's toxin will not create a triable issue of the level or duration of exposure, and require instead at least some medical or other expert confirmation of exposure.[20]

10.3.2 Substantial Contributing Factor

Under the substantive law of most jurisdictions plaintiff toxic tort claimant must prove that defendant's negligent or intentional conduct was more likely than not a cause of, or a ''substantial contributing factor'' in, plaintiff's particular injuries.[21] Proof that a toxin was ''more likely than not'' the cause of an injury or a disease is described as proof by a ''preponderance'' of the evidence. Failure of a claimant's expert to employ the term of art ''substantial contributing factor'' should not preclude submission of plaintiff's case to the jury where there is evidence that would permit a jury to conclude that defendant's products or conduct *contributed substantially* to the injury or disease.[22] Many courts have distinguished this so-called preponderance standard from ''scientific proof'', which usually requires a confidence level of 90 to 95%.[23]

Mere identification of a statistical correlation between exposure to the product and contacting the disease has been found to be inadequate in actions arising from exposure to asbestos.[24] It has, accordingly, been held in asbestos-related injury suits that plaintiff's proof of proximate cause must do more than equate risk exposure with causation.[25]

10.3.3 Proof of Exposure to Toxin

As a threshold matter, plaintiff must prove exposure to defendant's toxin. More specifically, for the toxic tort plaintiff to raise a triable issue of proximate cause, he or she must present ''direct or circumstantial evidence of at least one exposure to defendant's products.''[26] Additionally, in cumulative impact injuries, plaintiff generally is required to show the level and duration of exposure to a toxin through expert medical or scientific testimony.[27] In asbestos cases, for example, most courts have required proof ''that an injured plaintiff was exposed to a particular defendant's asbestos-containing product[.]''[28] For toxins whose characteristic is to leave trace elements in the victims body, failure to produce evidence of the toxin in a particular claimant imperils plaintiff's proof of proximate cause.[29]

Where plaintiff's proof of exposure is circumstantial, several courts have adopted a ''frequency, regularity and proximity'' test.[30] Plaintiff will not make out a prima facie showing of injurious exposure by merely showing that the toxin was delivered to or present in a workplace or other environment where plaintiff was present.[31] Even where plaintiff can prove that defendant's product was at the workplace, plaintiff's proof of exposure to the toxin will fail absent evidence that the product came close to plaintiff's particular work station, i.e., plaintiff must show that when the product was at the workplace, it was actually proximate to plaintiff's particular work station or stations.[32] Evidence of an employee's exposure to a toxin in the workplace may come from the employee's work records,[33] testimony of the employee's co-workers,[34] or evidence of the employer's ordinary procurement practices.[35]

Plaintiff may demonstrate defendant's responsibility for the toxin at issue by evidence eliminating other potential sources of the toxin, i.e., by proving that no operation other than defendant's could have released the toxin. Observing that "crumbs on the floor around the dining room table may be reasonably supposed to have fallen from the table," the Mississippi Supreme Court held that plaintiff in a suit against an herbicide manufacturer had satisfied its burden on a record revealing no release of such chemicals other than those from defendant's plant.[36] In contrast, defendant manufacturer should be permitted to introduce evidence showing that claimant was exposed to the toxic products of other nonparty producers.[37]

In pharmaceutical toxic tort actions where plaintiff alleges that *in utero* (prenatal) exposure to a drug caused newborn injury, defendant pharmaceutical manufacturer may successfully defend by demonstrating that the injury suffered was incompatible with the mother's evidence of when she was exposed to the drug. Illustrative is one suit against a manufacturer of Bendectin, in which defendant prevailed against the mother's claim that administration of the drug during the eighth week of pregnancy caused the infant's limb deformity by introduction of medical evidence that the onset of fetal limb deformity occurs before the eighth week.[38]

10.4 IDENTIFICATION OF DEFENDANT

10.4.1 Generally

Traditionally, plaintiff must show that the injured party was exposed to the toxic product of a particular defendant.[39] In the absence of exact records, such identification evidence can be circumstantial.[40] Even with the liberalization of proof requirements discussed below, plaintiffs cause of action against the manufacturer of a toxic substance will fail in the absence of evidence showing exposure to defendant's product, or, at the least, defendant's contribution to the risk or hazard to which plaintiff was exposed.[41]

Significantly, plaintiff need not bring all alleged joint tortfeasors together in a single action.[42] One court has suggested that a claimant's parallel suits against silica manufacturers and asbestos manufacturers arising from the worker's injury through contracting silicosis and asbestosis, respectively, may state a claim of liability for joint tortfeasors.[43]

10.4.2 Alternate Liability

If the injured party can bring before the court all defendants who breached a duty to plaintiff, in many jurisdictions plaintiff may sue all of the tortfeasors on a theory of "alternate" liability. The most influential early recitation of this theory was the 1948 California decision in *Summers v. Tice.*[44] The form of "alternate" liability later adopted by the Restatement, Second, of Torts § 433B(3) provides that where (1) plaintiff can prove injury was caused by one member of a group of tortfeasors; and (2) bring into court all potentially liable defendants; but (3) cannot demonstrate which particular defendant caused the injury, the burden of proof shifts to each of the joined defendants to exculpate himself.[45]

In one suit where the trial court applied alternate liability, it granted plaintiff leave to amend its complaint to state a cause of action in alternate liability against manufacturers of an antihemophilic blood product in *Poole v. Alpha Therapeutic Corp.*,[46] reasoning that "[s]ince plaintiffs have identified all of the defendants who could possibly have caused Poole to contract AIDS, this case presents the classic situation contemplated by the alternative liability doctrine."[47]

10.4.3 Concert of Action

One defendant may be jointly and severally liable for the injurious conduct of another where plaintiff can prove that defendant acted in concert with the other tortfeasor. Evidence of such concerted action most frequently takes the form of proof that defendants (1) acted in concert, or "pursuant to a common scheme;" or (2) knew of another defendant's breach of duty to plaintiff and assisted or encouraged that breach.[48] To prevail in a count of concerted action, plaintiff must do more than allege that multiple actors engaged in "identical negligent conduct".[49] Rather, examples of such conduct might be two or more manufacturers or other sellers having an "express agreement or tacit understanding . . . to prevent public awareness" of a product hazard, or even include an agreement, explicit or implicit, "to block governmental action" directed at remedying the risk.[50]

Most states have recognized a cause of action for "[j]oint liability for the tortious infliction of harm, whether intentional or not, against all persons acting in concert with the actual wrongdoer."[51] Under this theory, "one may be liable for the tortious conduct of another person by, inter alia, engaging in a tortious conduct in concert or pursuant to a common design with the other person or giving substantial assistance or encouragement to the other person with knowledge that the other's conduct constitutes a breach of duty."[52]

10.4.4 Market Share Liability

Relaxation of the common law rule of definitive defendant identification is premised upon employment, product, industry, injury latency, or recordkeeping circumstances that becloud the claimant's ability to specify the manufacturer of the injury-causing agent. In *Sindell v. Abbott Laboratories,*[53] a DES case, the California Supreme Court held that in view of plaintiff's inability to identify with certainty whose product caused plaintiff's personal injury, plaintiff could join the major producers of the hazardous substance in a single action. Upon satisfaction of plaintiff's burden of showing that the complained-of injuries resulted from exposure to a product produced by each of the defendants during the time period of the proved injurious exposure, the burden would shift to defendants to exculpate themselves. Defendants shown to have produced the product during the relevant time period were jointly responsible in damages to plaintiff in proportion to their market share in the sale and distribution of the product at the time of the injurious exposure. Defendants could exonerate themselves by producing positive proof that their product was not the product that caused the injury.

Variations on this "market share" approach have been reached by many courts in connection with DES suits. In most jurisdictions, however, suits involving toxins, and not DES, have found market share liability unsuitable to fair adjudication of plaintiffs' claims.[54] Courts in other jurisdictions have refused to adopt any version of market share liability irrespective of the nature of the toxic substance or the risk-creating circumstance.[55] Consistent with this, most courts have declined to apply market share liability to claims of asbestos-related tort actions. Principal reasons for differentiating DES claims from those involving asbestos center upon (1) defendant identification issues that differentiate DES from asbestos and (2) distinctive hazard and causation evidence pertaining to the two products.[56] Regarding the latter distinction, one court observed that "[u]nlike DES, which is fungible, asbestos fibers are of several varieties, used in varying quantities in the various products that contain asbestos, and each is different in its harmful effect."[57]

NOTES

1. *Robertson v. Allied Signal, Inc.*, 914 F.2d 360, Prod. Liab. Rep (CCH) ¶ 12,567 (3d Cir. 1990).
2. E.g., *In re Tire Workers Asbestos Litigation,* 125 F.R.D. 617 (E.D. Pa. 1989) ("To withstand summary judgment, plaintiffs must present some competent evidence that the defendant's product was present in the workplace, that the plaintiffs worked in the vicinity of defendant's product on a regular basis and inhaled fibers shed by defendant's product.").

3. E.g., *Malensky v. Mobay Chemical Corp.*, 104 Or. App. 165, 1990 (Or. App. 1990), a raspberry grower's action against the manufacturer of a pesticide, in which the grower and an entomologist testified that defendant's product altered the quality of the raspberry nectar and discouraged bees from pollinating plaintiff's raspberries. Compare *Thacker v. UNR Indus., Inc.*, 572 N.E. 2d 341, 213 Ill. App. 3d 38 (Ill. App. 1990) (evidence that defendant manufacturer supplied at most 3% of the asbestos present at the worksite during plaintiff's exposure, together with an absence of evidence that defendant's product was in proximity to plaintiff's work situs; held: insufficient to establish proximate cause under a substantial factor analysis); *McClelland v. Goodyear Tire & Rubber Co.*, 735 F. Supp. 172, Prod. Liab. Rep. (CCH) ¶ 12,619 at 37,746 (D. Md. 1990) (plaintiff workers alleging illness from exposure to workplace chemicals failed to adduce evidence "that would allow a reasonable fact-finder to conclude that it was more likely than not that any particular, identifiable negligent or intentional conduct by Goodyear was the cause of any of the plaintiffs' particular illnesses.").
4. E.g., *Reeves v. Armstrong World Indus.*, 569 S.E.2d 1307, (Fla. App. 1990) (insufficiency of evidence where plaintiff's medical expert testified only exposure to the manufacturer's products could conceivably have caused plaintiff's asbestosis); *Norman v. National Gypsum Co.*, 739 F. Supp. 1137, (E.D. Tenn. 1990) (statistical correlation between asbestos exposure and lung cancer insufficient to show causation; similarly, examining physician's willingness to give opinion that asbestos exposure was a causal factor in the plasterer's illness insufficient absent evidence of asbestos fiber in victim's body, and in light of heavy smoking habits of victim).
5. E.g., *Rohrbough v. Wyeth Laboratories, Inc.*, 916 F.2d 970, Prod. Liab. Rep. (CCH) ¶ 12,629 at 37,784 (4th Cir. 1990).
6. E.g., *Longmore v. Merrell Dow Pharmaceuticals, Inc.*, 737 F. Supp. 1117, 1120 (D. Idaho 1990) ("[T]he plaintiff need only prove that it is more probably true than not that the mother's ingestion of Bendectin caused David's Poland Syndrome."). See also *Rohrbough*, supra note 5. In that appeal of a DPT suit, the appellate court, affirming the trial court's grant of summary judgment for defendant vaccine manufacturer, stated that "[a]n essential element of plaintiff's cause of action is proof that defendant's vaccine caused plaintiff's injuries, and proof of causation must be by expert testimony.[citation] In addition, an expert's opinion as to proximate cause must be stated in terms of reasonable probability.[citation]" *Id.*, 916 F.2d at 972. Compare *Niklaus v. Vivadent, Inc.*, 767 F. Supp. 94, (M.D. Pa. 1991) (held: plaintiff dentist's claim that defendant's dental resin curing light caused eye damage unproven by plaintiff's expert whose testimony on causation did no more than match symptom with what the expert believed was a probable cause).
7. *Maddy v. Vulcan Materials Co.*, 737 F. Supp. 1528, 1533 (D. Kan. 1990) ("Expert opinion testimony must be grounded on reliable data, particularly in establishing causation in product liability toxic tort actions", citing *in re Agent Orange Product Liability Litigation,* 611 F. Supp. 1223, 1248–1250, 1261–1263 (D.N.Y. 1985), affirmed 818 F.2d 187 (2d Cir. 1987), *cert. denied* 487 U.S. 1234 (1988); *Turpin v. Merrell Dow Pharmaceuticals, Inc.*, 736 F. Supp.

737, 741 (E.D. Ky. 1990) ("[D]istrict judges 'faced with medical and epidemiological proof in subsequent toxic tort cases [should] be especially vigilant in scrutinizing the basis, reasoning, and statistical significance of studies presented by both sides." (quoting *Brock v. Merrell Dow Pharmaceuticals, Inc.*, 874 F.2d 307, *modified* 884 F.2d 166 (5th Cir. 1989), *cert. denied*, ____ U.S. ____, 110 S. Ct. 1511 (1990).

8. *Viterbo v. Dow Chemical Co.*, 646 F. Supp. 1420, 1424 (E.D. Tex. 1986), affirmed 826 F.2d 420 (5th Cir. 1987) ("rigorous examination" pertaining to satisfaction of Fed. R. Evid. 703 "is especially important in the toxic tort context, where presentation to the trier of theories of causation depends almost entirely upon expert testimony.")
9. E.g., *Allen v. Roddis Lumber and Veneer Co.*, 796 S.W.2d 758 (Tex. App. 1990) (expert testimony that formaldehyde levels in vapor emissions from defendant's furniture were below level that could cause respiratory problems).
10. *Burk v. Sage Products, Inc.*, 747 F. Supp. 285 (E.D. Pa. 1990).
11. Fed. R. Civ. Proc. 56(c).
12. Fed. R. Civ. Proc. 56(c) provides that summary judgment should be granted only where "there is no genuine issue as to any material fact and . . . the moving party is entitled to judgment as a matter of law." Summary judgment "must be entered against the nonmoving party, after adequate time for discovery, 'fails to make a showing sufficient to establish the existence of an element essential to that party's case, and on which that party will bear the burden of proof at trial.'" *Evenson v. Osmose Wood Preserving, Inc.*, 760 F. Supp. 1345, 1347 (S.D. Ind. 1990) (pesticide exposure involving chromated copper arsenate), citing *Celotex Corp. v. Catreff*, 477 U.S. 317, 322–323 (1986).
13. *Barwick v. Celotex Corp.*, 736 F.2d 946, 960, Prod. Liab. Rep. (CCH) ¶ 10,106 (4th Cir. 1984); accord, *Rohrbough v. Wyeth Laboratories, Inc.*, 916 F.2d 970, Prod. Liab. Rep. (CCH) ¶ 12,629 at 37,787 (4th Cir. 1990) ("Given the conflicts between Dr. Cox's affidavit and his deposition testimony, the district court was left not with a genuine issue of material fact, but with trying to determine which of the several conflicting versions of Dr. Cox's testimony was correct."[citation]).
14. 895 F.2d 1463, Prod. Liab. Rep. (CCH) ¶ 12,374 (D.C. Cir. 1990).
15. *Id.*, Prod. Liab. Rep.(CCH) ¶ 12,374 at 36,601 ("Taking these statements together, a reasonable juror could conclude that the only medication that fits the medical and physical description of the drug which Mrs. Shields took while pregnant with appellant was DES.").
16. *Abbott v. Babcock & Wilcox*, 905 F.2d 201, Prod. Liab. Rep. (CCH) ¶ 12,471 at 36,996 (8th Cir. 1990) ("[W]e note that there was evidence that Mr. Abbott was exposed to Carey products at various times and at intervals for possibly as long as ten years. There was also evidence that he was exposed for considerably less time. The hypothetical asked of Dr. Tuteur reflects that one witness indicated exposure to ten years and others indicated other lengths of time. The fact that the evidence was disputed on this point does not force the conclusion that the hypothetical was based on facts not in evidence. We do not resolve factual disputes on appeal, and all hypotheticals are based on at least some

disputed facts. In some sense, this is an issue of credibility which must be left to the jury[.]'').

17. *Benshoff v. National Gypsum Co.*, 761 F. Supp. 671 (D. Ariz. 1991) (consolidated asbestos cases, Arizona law).
18. *Benshoff v. National Gypsum Co.*, 761 F. Supp. 677, 679 (D. Ariz. 1991) (citation omitted) (asbestos-related injury).
19. E.g., *Maddy v. Vulcan Materials, Inc.*, 737 F. Supp. 1528, 1533 (D. Kan. 1990) (''Dr. Pollard was unable to isolate other potential causes of Maddy's respiratory illness, admitting that her illness may have been due to her smoking or to other environmental factors.'').
20. *Maddy v. Vulcan Materials Co.*, 737 F. Supp. 1528, 1533 (D. Kan. 1990) ('' . . . Maddy presented no scientific or expert evidence that she was in fact exposed to these chemicals . . . [there was] no independent attempt to verify the existence of the exposure. In cases claiming personal injury from exposure to toxic substances, it is essential that the plaintiff demonstrate that she was, in fact, exposed to harmful levels of such substances. [Citations] In the present case, there is no scientific evidence indicating the level or duration of Maddy's exposure to specific toxins.''). See also *Celotex Corp. v. Catreff*, 477 U.S. 317, 319–320 (1986); *Davis v. DuPont*, 729 F. Supp. 652, 655 (E.D. Ark. 1989).
21. E.g., *McClelland v. Goodyear Tire & Rubber Co.*, 735 F. Supp. 172, Prod. Liab. Rep. (CCH) ¶ 12,619 (D. Md. 1990) (adopting preponderance of evidence test for causation in plaintiff workers' claim against tire manufacturer for allegedly injurious exposure to workplace toxins).
22. *Abbott v. Babcock & Wilcox*, 905 F.2d 201, Prod. Liab. Rep. (CCH) ¶ 12,471 at 36,996–36,997 (8th Cir. 1990). In that suit, brought under Missouri law and arising from plaintiff's occupational exposure to asbestos insulation, the court states: ''[W]here Dr. Tuteur said only that Carey products were a contributing factor in Mr. Abbott's death, there is room for reasonable minds to conclude that Carey products were not a *substantially* contributing factor in his death. But there is room to conclude that they were. That is a point best argued to a jury, not an appellate court.'' *Id.*, Prod. Liab. Rep. (CCH) ¶ 12,471 at 36,997.
23. Rothman, *Modern Epidemiology*, 119 (1986). E.g., *Wells v. Ortho Pharmaceutical Corp.*, 788 F.2d 741, 745, Prod. Liab. Rep. (CCH) ¶ 10,997 (11th Cir. 1986) (''a distinction exists between legal sufficiency and scientific certainty.'').
24. E.g., *Norman v. National Gypsum Co.*, 739 F. Supp. 1137, (E.D. Tenn. 1990).
25. See *Landrigan v. Celotex Corp.*, 243 N.J. Super. 449, 579 N.E.2d 1268 (N.J. Sup. App. Div. 1990) (trial court's refusal to permit jury to credit medical witness' opinion justified in part on basis that opinion seemingly equated risk exposure with causation).
26. *Bensoff v. National Gypsum Co.*, 761 F. Supp. 677, 680 (D. Ariz. 1991) (consolidated asbestos actions).
27. E.g., *Rubanick v. Witco Chemical Corp.*, 543 A.2d 975 (N.J. Law Div. 1988), rev'd, 576 A.2d 4 (N.J. App. Div. 1990).
28. *Benshoff v. National Gypsum Co.*, 761 F. Supp. 677 (D. Ariz. 1991) citing, but not following, proximate cause standard of *Cohrmann v. Pittsburgh Corning*

Corp., 782 F.2d 1156 (4th Cir. 1986) (Maryland law) and *Menore v. Celotex Corp.*, 861 F.2d 1453 (10th Cir. 1988) (Nebraska law). In *Benshoff* the court held that plaintiff's simple showing that asbestos was present at plaintiff's workplace fell short of identifying "the manufacturer, in what products the asbestos was found, and whether the fibers of those products were in a disturbed state while plaintiff was present. . . . There was nothing in the record that placed Mr. Benshoff at the locations where the OCF products were being installed or disturbed so that the fibers from the product would be airborne." 761 F. Supp. at 679–680. Compare *Eagle-Picher Industries, Inc. v. Balbos*, 84 Md. App. 10, Prod. Liab. Rep. (CCH) ¶ 12,562 (Md. Spec. App. 1989) (evidence of exposure to manufacturer's asbestos products sufficient).

29. E.g., *Norman v. National Gypsum Co.*, 739 F. Supp. 1137, (E.D. Tenn. 1990) (absence of evidence of asbestos in plasterer victim's lungs, coupled with evidence of heavy smoking habits of victim).

30. E.g., *Lohrman v. Pittsburgh Corning Corp.*, 782 F.2d 1156, 1162, Prod. Liab. Rep. (CCH) ¶ 10,928 (4th Cir. 1986) ("To support a reasonable inference of substantial causation from circumstantial evidence, there must be evidence of exposure to a specific product on a regular basis over some extended period of time in proximity to where plaintiff actually worked.").

31. *Benshoff v. National Gypsum Co.*, 761 F. Supp. 677, 679 (D. Ariz. 1991) ("Mr. Benshoff's exposure to OCF products was limited. He cannot recall any jobsite where he could identify any asbestos-bearing product manufactured by defendant OCF. He was employed at several jobsites where OCF products were located or installed, but even this circumstantial evidence by way of co-worker testimony merely established that asbestos was present."). See also *Hyde v. Owens-Corning Fiberglas Corp.*, 751 F. Supp. 832, 833 (D. Ariz. 1990) [("The fact that a defendant's asbestos-bearing product was supplied to a work-site is insufficient to establish liability.") (citing *Blackston v. Shook and Fletcher Insulation Co.*, 764 F. Supp. 1480, 1481 (11th Cir. 1985)].

32. E.g., *Thacker v. UNR Indus., Inc.*, 572 N.E.2d 341, 213 Ill. App.3d 38, (Ill. App. 1990). In Thacker, plaintiff's evidentiary problems were compounded by evidence that defendant provided at most 3% of the asbestos present at the workplace.

33. *Abbott v. Babcock & Wilcox*, 905 F.2d 201, Prod. Liab. Rep. (CCH) ¶ 12,471 (8th Cir. 1990).

34. *Abbott v. Babcock & Wilcox*, supra, Prod. Liab. Rep. (CCH) ¶ 12,471 at 36,996 ("The evidence of Mr. Abbott's contact with asbestos products came from two primary sources: his own work records which he logged in journals and the testimony of his co-workers. Without reciting the evidence, but viewing it in the Abbotts' favor, we note that it revealed that Mr. Abbott had contact with asbestos products throughout his career as an insulator, and in particular that he had contact with Carey's products on different occasions totalling somewhere between three and one half years to ten years."); *Taylor v. Celotex*, 574 A.2d 1084 (Pa. Super. 1990) (asbestos action; testimony of co-workers that plaintiff had worked on specific vessels at the times the manufacturer's products were in use).

35. E.g., *Mason v. Texaco, Inc.*, 741 F. Supp. 2472 (D. Kan. 1990). In *Mason* the court held that the product identification component of a Coast Guard worker's claim that exposure to the manufacturer's benzene product caused his leukemia was amply supported by evidence of the Coast Guard's practice of not reordering the product until its supply was low, permitting conclusion that claimant was exposed only to defendant's product.
36. *Shutes v. Platte Chemical Co.*, 564 So. 2d 1382 (Miss. 1990) (herbicide manufacturer of linuron, claimed to be cause of damage to vegetation on plaintiff's property, shown to be only manufacturer of product within 200-mile radius of damage).
37. See *Laney v. Celotex Corp.*, 901 F.2d 1319, (6th Cir. 1990).
38. *Obiago v. Merrell-National Laboratory, Inc.*, 560 S.2d 625 (La. App. 1990).
39. E.g., *Celotex Corp. v. Tate*, 979 S.W.2d 197, Prod. Liab. Rep. (CCH) ¶ 12,673 (Tex. App. 1990) (action by estate of worker who dies of pleural mesothelioma allegedly resulting from prolonged workplace exposure to defendant's asbestos products, used as one component of kalite, the soundproofing material plaintiff's decedent prepared and handled).
40. *Id.*, *Celotex Corp. v. Tate*, 797 S.W.2d 197, Prod. Liab. Rep. (CCH) ¶ 12,673 (Tex. App. 1990) (circumstantial evidence that defendant's asbestos products were among those used by plaintiff's decedent included testimony from plant manager that defendant's product was used in the manufacture of soundproofing during the period of decedent's employment, and from a former co-worker who recalled the workplace delivery of bags of asbestos marked with defendant's corporate name).
41. E.g., *White v. Celotex Corp.*, 907 F.2d 104, Prod. Liab. Rep. (CCH) ¶ 12,546 (9th Cir. 1990) ("The Whites proffered no evidence that Edward White had been exposed to any particular product produced or distributed by the Defendants. The fact that defendants were making or marketing asbestos between 1957 and 1983 did not constitute evidence that they had done so in relation to the asbestos in any building with which Edward White may have come into contact. For all that the Whites could show, even if their affidavits were given full weight, any asbestos effecting Edward White had been installed before 1957 and as to this period the Whites proffered no proof as to the activities of the Defendants. Hence the Whites had no case against the Defendants.") *Id.*, 907 F.2d at 106. See *Case v. Fibreboard Corp.*, 743 P.2d 1062, 1066–1067 (Okla. 1987) ("[T]he public policy favoring recovery on the part of an innocent plaintiff does not justify the abrogation of the rights of a potential defendant to have a causative link proved between the defendant's specific tortious acts and the plaintiff's injuries where there is a lack of circumstances which would insure that there was a significant probability that those acts were related to the injury.").
42. *Temple v. Synthes Corp. Ltd.*, _____ U.S. _____, 111 S. Ct. 315 (U.S. 1990) (in action against manufacturer of "plate and screw" device implanted in spine, held: hospital and physician performing the procedure were potential joint tortfeasors, but Fed. R. Civ. Proc. 19, pertaining to indispensable parties, does not change the longstanding rule that plaintiff need not join all joint tortfeasors in a single suit).

43. *Arabie v. Northwest Min. Corp.*, 567 So. 2d 783 (La. App. 1990). The court added that the possibility of joint-tortfeasor liability could affect the calculation of the prescriptive period for the later-filed suit against the silica manufacturers. See discussion of Statutes of Limitations infra in Chapter 15.
44. 33 Cal. 2d 80, 199 P.2d 1 (1948).
45. Restatement, Second, Torts § 433(B), which provides that "[w]here the conduct of two or more actors is tortious, and it is proved that the harm has been caused to the plaintiff by only one of them, but there is uncertainty as to which one has caused it, the burden is upon each such actor to prove that he has not caused the harm."
46. 696 F. Supp. 351, Prod. Liab. Rep. (CCH) ¶ 12,047 (N.D. Ill. 1988).
47. *Id.*, Prod. Liab. Rep. (CCH) ¶ 12,047 at 34,962.
48. "For harm resulting to a third person from the tortious conduct of another, one is subject to liability if he (a) does a tortious act in concert with the other or pursuant to a common design with him; (b) knows that the other's conduct constitutes a breach of duty and gives substantial assistance or encouragement to the other to so conduct himself, or (c) gives substantial assistance to the other to accomplishing a tortious result and his own conduct, separately considered, constitutes a breach of duty to the third person." Restatement, Second, Torts § 876.
49. *Poole v. Alpha Therapeutic Corp.*, 696 F. Supp. 351, Prod. Liab. Rep. (CCH) ¶ 12,047 at 34,961 (N.D. Ill. 1988) (action against several manufacturers of antihemophilic blood product; held: plaintiff would not be granted leave to amend complaint to include claims of concerted action liability or market share liability, although plaintiff could plead alternate liability under Restatement, Second, Torts § 433B(3)).
50. *Rastelli v. Goodyear Tire & Rubber Co.*, 565 N.Y.S.2d 889 (App. Div. 1991) (denying manufacturer's motions for summary judgment on plaintiff's concert of action claims against manufacturers of multipiece tire rims, the court stating its recognition of evidence plaintiffs might develop during discovery that could, if admitted into evidence at trial, support concert of action liability against even those manufacturers capable to proving that their particular products were not involved in plaintiff's injury).
51. E.g., *Rastelli v. Goodyear Tire & Rubber Co.*, 565 N.Y.S.2d 889 (App. Div. 1991) (suit against multiple defendants engaged in the manufacture of multipiece tire rims).
52. Rastelli, *id.*, 565 N.Y.S.2d 889 (App. Div. 1991) (recognizing plaintiff's potential concert of action claim against multipiece rim manufacturer for failure to warn, and postsale failure to warn of risks, even where evidence demonstrated that the product of a particular defendant was not involved in the harm to plaintiff).
53. 26 Cal. 3d 588, 163 Cal. Rptr. 132, 607 P.2d 924 (1980), *cert. denied* 449 U.S. 912 (1980).
54. E.g., *Leng v. Celotex Corp.*, 196 Ill. App. 3d 647, 554 N.E.2d 468, Prod. Liab. Rep (CCH) ¶ 12,472 (Ill. App. 1990). In that action, the appellate court found support for denying an estates' modified market share liability action against several asbestos manufacturers in its recognition that, based upon the

averments in plaintiff's complaint, "plaintiff has some ability to identify certain manufacturers who allegedly caused decedent's injury. Such identification would, of course, preclude the use of market share liability." Accord, *Copeland v. Celotex Corp.*, 471 So. 2d 533 (Fla. 1985); *Prelick v. Johns-Manville Corp.*, 531 F. Supp. 96 (W.D. Pa. 1982).

55. E.g., *McClelland v. Goodyear Tire & Rubber Co.*, 735 F. Supp. 172, Prod. Liab. Rep. (CCH) ¶ 12,619 at 37,745–13,746 (D. Md. 1990) (worker claimants alleging illness due to exposure to toxic chemicals; "Maryland has not adopted the 'market share' theory of product liability, of which the plaintiffs here assert a microcosmic variant.").

56. E.g., *Leng v. Celotex Corp.*, 196 Ill. App. 3d 647, 554 N.E.2d 468, Prod. Liab. Rep. (CCH) ¶ 12,472 (Ill. App. 1990) (With asbestos-related illness, "the injury caused, unlike that caused by DES, does not arise at a specific time. Asbestos continues to cause injury throughout the time of its use and exposure." *Id.*, Prod. Liab. Rep. (CCH) ¶ 12,472 at 36,999.).

57. *In re Related Asbestos Cases*, 543 F. Supp. 1152, 1158 (N.D. Cal. 1982).

CHAPTER 11

Nonexpert Proof in a Toxic Tort Case

CONTENTS

Nonexpert Proof in a Toxic Tort Case

11.1 GENERALLY

In a toxic tort claim, plaintiff must prove that (1) the toxic substance was unreasonably dangerous; (2) plaintiff suffered personal injury or property damage; (3) defendant's product, process, or activity was a direct cause of the injury or loss; and (4) that defendant was responsible for the condition of the product, or the risks created by the process or activity.

Prior to trial, major discovery initiatives are undertaken by plaintiff and defendants alike. Pursuant to federal and state rules of discovery, the parties may exchange voluminous amounts of information. The trial judge has substantial discretion to appoint a master to hear and resolve pretrial discovery matters. Such discretion has limitations, however, and assignment of pretrial matters to a master may be held as an abuse of discretion without a showing that the suit is somehow exceptional in the anticipated breadth or complexity of discovery matters.[1]

Counsel may be disqualified from representing a party to a toxic tort suit upon a showing that in some prior capacity he or she had access to the confidential materials of an adversary party. Efforts to disqualify counsel for a party because of claimed earlier access to confidential, design, formulation, production, or other information will fail without proof of the actual content of the confidential information claimed to have been compromised.[2]

11.2 TIME OF EXISTENCE OF DEFECT

In any tort claim, plaintiff must show that the product was in an unreasonably dangerous condition at the time it left the possession of the manufacturer or other marketer. Factors that support the seller's claim that the product was not in a defective condition at the time of sale can include, among other considerations: (1) the passage of time from the sale to the claimed injurious exposure; (2) the alteration of the product; or (3) proof of uneventful product use prior to the claimed injurious exposure.

These considerations came into play in *Sutton v. Major Products Co.*,[3] a claim brought by a grocery store employee who alleged injury from the exposure to and inhalation of fumes and vapors from a jar of potato whitener. Affirming the trial court's entry of judgment against plaintiff, the North Carolina appeals court noted evidence showing that by the time of the employee's exposure to the product, approximately one third of the potato whitener jar's contents had been used without incident. The court observed additionally that plaintiff had failed to produce any rebuttal evidence that the product was unreasonably dangerous when it was distributed initially.

11.3 OTHER ACCIDENTS OR INJURIES

Upon a showing of sufficient similarity of product and circumstances of administration, a plaintiff claiming injury from use of a defendant's product may present evidence of similar accidents or adverse reactions suffered by others who were exposed to the product or substance. Such evidence will not, however, be admitted where plaintiff fails to show such substantial similarity.[4]

Illustrative of the need for sufficient similarity is the decision in *Nasios v. Pennwalt Corp.*,[5] a suit in which the claimant claimed that her partial paralysis was caused by epidural injections of defendant's anaesthetic during labor: a trial court disallowed evidence of injuries suffered by other patients who received injections of the same anaesthetic in the subarachnoid space of the spinal column. Measuring the probative value of plaintiff's proffer against its prejudicial effect, pursuant to Federal Rule of Evidence 403, the court ruled that its probative value was simply too weak to warrant admissibility.[6] A comparable result was reached in a suit brought against the manufacturer of chymopapain by a patient who became permanently paralyzed following injection of the drug for chronic back pain.[7] There the state supreme court affirmed the trial court's preclusion from evidence prior alleged incidents of adverse reactions to the drug that resulted in paresis, as distinct from paraplegia.[8]

Similarly, in *Marsee v. United States Tobacco Co.*,[9] plaintiff, as representative of a user of defendant's tobacco snuff products, claimed that her son died of oral cancer caused by this product use. At trial, plaintiff sought unsuccessfully to introduce evidence of other victims of oral cancer who had used snuff products. The appeals court sustained the trial court's refusal to admit this evidence, using as one basis its conclusion that oral cancer is a relatively common form of cancer and is generally thought to be caused by many factors.

The absence of claims of injury or damage from prior use of or exposure to a product is evidence that the product is not unreasonably dangerous.[10] For example, plaintiff asserting injury caused by an allergic reaction to defendant's product must identify with particularity prior incidents by which the manufacturer knew or had reason to know that some consumers might be allergic to its products. Thus, in *Burlison v. Warner-Lambert Co.*,[11] plaintiff's failure to show prior incidents sufficiently similar in nature to put the manufacturer on notice of potential product risks has defeated claims alleging allergic reactions to cough drops.[12]

In a nondiscovery context, an employer may enjoy a limited privilege to disclose otherwise private facts concerning an employee's illness where such disclosure serves directly to allay employee concerns that the illness was related to workplace hazards.[13]

11.4 OTHER EXPOSURE

The toxic tort defendant should be permitted to introduce evidence of a claimant's exposure to the toxic products of another producer.[14] Hence, a defendant asbestos manufacturer defending a claim that a former pipefitter sustained asbestos-related lung disease due to exposure to defendant's products should be permitted to adduce proof that it is equally probable that plaintiff's lung disease was caused by years of using tobacco products. Likewise, a chemical solvent manufacturer's defense of a claim brought by a plaintiff claiming that his liver cancer was proximately caused by exposure to defendant's product should be permitted to present the court or the jury with evidence of plaintiff's consumption of alcoholic beverages, and the risks of liver cancer occasioned by that exposure.

11.5 PUBLISHED REPORTS

Where a report published by a governmental agency does not reflect the findings of a government unit on the toxicity or hazard at issue, courts have not permitted their introduction into evidence under Fed. R. Evid. 803(8). Thus in one tobacco products case, it was held that the trial court correctly excluded from evidence two agency reports. One report, described as "a review and critique of the literature on the causation of cancer, conducted at a conference attended by scientists from various countries[,]" and the second "prepared by a panel of scientists and non-scientists after a conference at which invited speakers presented papers concerning different aspects of smokeless tobacco." Neither report, the federal appeals court concluded, "reflects the findings of a governmental agency authorized by law to report on the adverse health effects of smokeless tobacco."[15]

11.6 DOCUMENTARY SUMMARIES

Large toxic tort litigation often prompts efforts to introduce evidence of document summaries. The federal rules and their state counterparts permit introduction of summaries upon proponent's showing that there is sufficient factual basis for admitting them, and that "possible prejudice or confusion does not outweigh their usefulness in clarifying the evidence."[16] Accordingly, in one asbestos-related claim on behalf of a subcontractor's employee, a summary witness was permitted to testify as to the contents of boxes of sales-related documents as such documents might pertain to whether the products of any given supplier were involved.[17]

11.7 POSTINCIDENT REMEDIAL MEASURES

Generally, in tort cases plaintiff is not permitted to introduce evidence that following plaintiff's injury, the defendant took measures to modify or eliminate the claimed risk. The logic of precluding this evidence is that postincident remedial measures have no reliable correlation to the presence of an unreasonably dangerous product quality, inasmuch as the defendant's remediation may not be motivated by the accident at all, or may simply reflect an abundance of caution, i.e., apprehension of tort liability might prompt some manufacturers to make safer a product that is already reasonably safe. This limitation on evidence is codified for federal courts at Federal Rule of Evidence 407.

Although the common law evidence limitation of Fed. R. Evid. 407 applies, by its terms, only to suits in negligence, some courts have concluded that the limitations should also apply in actions brought in strict tort liability.[18] Illustrative is *Fish v. Georgia-Pacific Corp.*,[19] the appeal of a personal injury and property damage claim arising from the installation of defendant's particleboard floor underlayment containing adhesive made from formaldehyde resins. The particleboard at issue here was installed at plaintiff's premises in 1977 and 1978. In a later suit, plaintiff claimed, among other things, that Georgia-Pacific should have warned of the risks of household construction with products containing formaldehyde. The trial court admitted into evidence plaintiff's showing that in 1983 defendant commenced providing warnings with the subject products, reasoning that evidence of this later warning was relevant as tending to prove that it would have been feasible for Georgia-Pacific to provide warnings before, i.e., at the time plaintiff purchased the products. Reversing the trial court's decision, the appeals court observed that the 1983 warning should not have been admitted on the issue of feasibility of providing warnings because Georgia-Pacific had conceded that warnings were feasible. The court continued: "We have held that FRE 407 applies in a strict products liability action to bar evidence of a remedial design change in a product where the change was made subsequent to the product's injuring the plaintiff and the evidence is offered on the issue of the manufacturer's culpable conduct. . . . [The] 1983 warning represents [such] a remedial measure", and, the appeals court decided, its admission into evidence was a reversible error.[20]

Similarly, in *Nasios v. Pennwalt Corp.*[21] the defendant manufacturer of an anaesthetic successfully precluded plaintiff's introduction of evidence of product changes postdating the incident forming the basis of plaintiff's claim. The trial court concluded: "Evidence of subsequent remedial measures, whether in the formulation of the product or in warnings attached to the product, are clearly inadmissible as direct evidence of liability under FRE 407.''[22]

Evidence of postincident remedial measures may be admitted, however, where it is logically probative of issues other than defendant's negligence or culpable conduct,[23] such as, for example, the ownership or control of a company, the identity of a product's manufacturer, or the feasibility of precautionary measures where defendant contests that there was no feasible way to produce a less hazardous product. It may also be considered by the court where presented by a nonmoving party in response to an opponent's dispositive motion.[24]

Admissibility of postincident remedial measures has been litigated vigorously in actions claiming personal injury or property damage caused by the use of the insecticide chlordane.[25]

11.8 COMPLIANCE WITH A STATUTE OR REGULATION

Generally, a defendant's compliance with standards set by a public agency will not preclude a finding that the product was unreasonably dangerous or that the manufacturer failed to provide adequate warnings.[26] For example, *Shorter v. Champion Home Builders Co.*[27] was a family's toxic tort claim against the builder of their home, alleging that the particleboard flooring of the dwelling ''contained dangerous levels of formaldehyde which caused severe respiratory and allergic reactions.'' Champion responded that the compliance of its construction materials with the applicable standards under the National Manufactured Home Construction and Safety Standards Act of 1974 (NMHCSSA)[28] constituted a complete defense to plaintiff's claims. The federal trial court disagreed, finding that defendant's compliance with the NMHCSSA standards could ''be treated as a piece of evidence going toward the issue of defective design[,]'' but did not preclude a finding that the manufacturer should have undertaken additional warning, labeling, or design initiatives. Other decisions have held that a chemical manufacturer's compliance with pertinent Environmental Protection Agency standards should not be given ''conclusive weight'' in a toxic tort claim.[29]

11.9 FAILURE TO COMPLY WITH A STATUTE OR REGULATION

Violation of state or federal toxicity disclosure or other safety regulations can be found to be a breach of the seller's duty to plaintiff. The appeals court in *Byrne v. SCM Corp.*,[30] an epoxy paint fume exposure suit, stated the general rule that ''[i]n order to recover for a defendant's violation of a statute or rule designed to protect human life or property, a plaintiff must show: (1) a violation proximately caused the injury; (2) plaintiff belonged to the class of persons whom the rule was intended to protect from injury; and

(3) the kind of injury suffered by plaintiff was the kind of injury which the rule sought to prevent.''[31] In that suit, defendant distributor had failed to comply with an Illinois law requiring that toxicity and cautionary information be provided purchasers of listed toxic substances. The court found that the defendant's failure to do so was a breach of the duty to warn the ultimate user, in this case a painter, of a known danger, and approved the trial court's instruction to the jury that the seller's failure to provide the required material safety data sheet in a timely manner rendered the product unreasonably dangerous.

11.10 DISCOVERY

The standard for discovery under the Federal Rules of Civil Procedure is that a requesting party is entitled only to such documentary or other evidence that would be admissible in court, or which is likely to lead to the discovery of admissible evidence. Accordingly, courts have regularly permitted discovery of documentary or other evidence without the necessity of determining at that time whether the discovered material will ultimately be admissible in trial, i.e., the party need only show that the requested material, if not itself admissible, will probably lead to discovery of other evidence that is admissible.[32]

An example of the practical operation of this liberal discovery rule might arise in a plaintiff's discovery request of all documents held by a defendant manufacturer that pertain to risk analyses made of a particular product. Defendant might have in its files risk analyses it received from the Consumer Product Safety Commission (CPSC) in regards to a filing under that agency's ''substantial product hazard'' reporting requirements.[33] While it would seem, at first glance, that these risk analyses should be freely discoverable, defendant would argue that CPSC regulations provide specifically that Commission findings shall not be introducible as evidence in any tort claim filed concerning the product that is the subject of a ''substantial product hazard'' evaluation. Yet the CPSC documents, even if not introducible into evidence, should be made available to plaintiff pursuant to this discovery request inasmuch as they may lead to the discovery of other, admissible evidence. Consistent with this rationale, in suits against manufacturers of medical devices, discovery has been permitted of Food and Drug Administration records and documents.[34]

NOTES

1. *Simpson v. Canales*, 806 S.W.2d 802 (Tex. 1991).
2. E.g., *Graham v. Wyeth Laboratories*, 760 F. Supp. 1451 (D. Kan. 1991) (court's refusal to disqualify plaintiff's counsel in DPT suit on the basis of the claim that an earlier engagement by an associate of counsel, as a member of the law firm representing defendant pharmaceutical manufacturer in another matter, gave access to privileged information).
3. 372 S.E.2d 897 (N.C. App. 1988).
4. E.g., *Renfro v. Black*, 52 Ohio St. 3d. 27, 1990 WL 79051 (Ohio App. 1990) (plaintiff's evidence of paraparesis and other neurological reactions to administration of Chymodiactin properly excluded for absence of proof that the other incidents were accompanied by circumstances substantially similar to her treatment and paralysis).
5. _____ F. Supp. _____, Prod. Liab. Rep. (CCH) ¶ 12,479 (D. Md. 1990).
6. *Id.*, Prod. Liab. Rep. (CCH) ¶ 12,479 at 37,045 (D. Md. 1990) ("[D]efendant presents a compelling argument that evidence of injuries resulting from subarachnoid administration of Nesacaine-CE is not "substantially similar" to evidence of injuries due to epidural injection of the anaesthetic.[Citation] Such evidence, therefore, should not be submitted to the jury; its probative force is simply too weak.").
7. *Renfro v. Black*, 52 Ohio St. 3d 27, Prod. Liab. Rep. (CCH) ¶ 12,524 (Ohio 1990).
8. *Id.*, ¶ 12,524 at 37,519–37,520.
9. _____ F.2d _____, Prod. Liab. Rep. (CCH) ¶ 12,023 (10th Cir. 1989).
10. *Sutton v. Major Products Co.*, 372 N.E.2d 897 (N.C. App. 1988) (affidavit of defendant distributor of potato whitener tended to show that "this jar of potato whitener had been opened and reopened and used several times before the accident and without any injury." *Id.*, 372 N.E.2d at 899.).
11. 842 F.2d 991, Prod. Liab. Rep. (CCH) ¶ 11,713 (8th Cir. 1988).
12. *Id.*, 842 F.2d 991, Prod. Liab. Rep. (CCH) ¶ 11,713 at 33,403 ("The district court did not err in concluding that there was insufficient evidence to support a verdict for the plaintiff. Burlison's expert witness . . . testified that he had reviewed the medical literature, but could not point to any reported case of a Hall's tablet causing an allergic or anaphylactic reaction." *Id.* at 33,403.).
13. E.g., *Young v. Jackson, Miss.* No 89-CA-0249, 1990 Toxic Law Reporter (BNA) 822 (November 28, 1990), in which the defendant nuclear power company revealed to employees that a woman's sudden workplace illness was related to an earlier hysterectomy to remove a fibroid tumor, and not to radiation exposure. The court concluded: "Prudence counseled full disclosure. What [defendant's agents] did was within the realm of good judgment such that we should not judicially second guess them."
14. E.g., *Laney v. Celotex Corp.*, 901 F.22d 1319 (6th Cir. 1990) (evidence of plaintiff's exposure to asbestos products of other manufacturers).
15. *Marsee v. United States Tobacco Co.*, _____ F.2d _____, Prod. Liab. Rep. (CCH) ¶ 12,023 (10th Cir. 1989).
16. Weinstein and Berger, Weinstein's Evidence Manual ¶ 9.05[02] at 9-32 (Matthew Bender 1991) (interpreting Fed. Rule Evid. 1006).

17. See *Fagiola v. National Gypsum Co.*, AC & S, Inc., 906 F.2d 53 (2d Cir. 1990).
18. See *De Luryea v. Winthrop Labs.*, 697 F.2d 222, Prod. Liab. Rep. (CCH) ¶ 9476 (8th Cir. 1983). See also Madden, "Admissibility of Post-Incident Remedial Measures: A Pattern Emerges," 5 J. Prod. Liab. 1 (1982).
19. 779 F.2d 836, Prod. Liab. Rep. (CCH) ¶ 10,904 (2d Cir. 1985).
20. *Id.*, 779 F.2d at 839.
21. _____ F. Supp. _____, Prod. Liab. Rep. (CCH) ¶ 12,479 at 37,046 (D. Md. 1990).
22. *Id.*, ¶ 12,479 at 37,046. The court added: "Pennwalt does not contest ownership, control, feasibility of precautionary measures, or any other matter that would render subsequent remedial measures pertinent to the litigation."
23. *Davis v. Dupont*, 729 F. Supp. 652, Prod. Liab. Rep. (CCH) ¶ 12,516 (E.D. Ark. 1990) involved a worker's claim that his hepatitis resulted from exposure to the toxic fumes of several unspecified paint products. Plaintiff sought to introduce a Material Safety Data Sheet (MSDS), published by defendant DuPont for its product Centari Acrylic Enamel, and which conceded that "recurrent exposure to [the toluene and xylene within that product] may result in liver and kidney injury." The MSDS was published after plaintiff's last possible exposure to the product. Defendant PPG Industries argued that the MSDS should not be admitted as it constituted evidence of a subsequent remedial measure. The trial court found its admission permissible, however, reasoning: "[T]he evidence is being presented not necessarily to prove negligence, but rather as an admission of DuPont that these chemicals have been linked with the type of injury suffered by Mr. Davis." *Id.* at 37,210.
24. See *Celotex Corp. v. Catrett*, 477 U.S. 317, 324 (U.S. 1986) (nonmoving party need not "produce evidence in a form that would be admissible at trial.").
25. E.g., *Dine v. Western Exterminating Company*, [not in F. Supp.], 1988 WL 28241 (D.D.C. 1988) (plaintiffs sought to introduce evidence of an 1987 agreement between Velsicol Chemical Corporation, manufacturers of chlordane and heptachlor, and the Environmental Protection Agency (EPA) to limit the registered uses of the chemicals; held: evidence of the agreement between the EPA and Velsicol was admissible, but that the agreement is not probative of the toxicity or carcinogenicity of chlordane and heptachlor and a jury could not draw such an inference); *Villari v. Terminix International*, 692 F. Supp. 568 (E.D. Pa. 1988) (evidence that termiticides containing chlordane and other chemicals had been taken off the market was also inadmissible to show that the termiticide was hazardous); *Rabb v. Orkin Exterminating Co.*, 677 F. Supp. 424 (1987) (refusing admission of evidence that exterminator discontinued use of termiticide and that manufacturer temporarily took termicitide off the market).
26. *Wyeth Ayerset Laboratories v. Graham*, _____ F.2d _____, 5 Toxic Law Rptr. (BNA) 176, *cert. denied*, 1990 Toxic Law Reporter 821 (November 28, 1990) (No. 90-476) (design defect and failure to warn claim brought against manufacturer of DPT vaccination).
27. _____ F. Supp. _____, 1991 WL 212776 (N.D. Ohio 1991).
28. 42 U.S.C. 5401-5427.
29. *Ferebee v. Chevron Chemical Co.*, 736 F.2d 1529, 1540 (D.C. Cir. 1984) (absent preemption, EPA standard "may be taken into account by the jury . . . but . . . the jury need not give that determination conclusive weight.").

30. 538 N.E.2d 796, Prod. Liab. Rep.(CCH) ¶ 12,170 (Ill. App. 1989).
31. *Id.*, 538 N.E.2d 796 at 813.
32. E.g., *Phillips v. Medtronic, Inc.*, [not in F. Supp.], 1990 WL 58440 (D. Kan. 1990) (discovery of manufacturer's postincident remedial measures).
33. 15 U.S.C. 2064 *et seq.*
34. *Phillips v. Medtronic, Inc.*, [not in F. Supp.], 1990 WL 58440 (D. Kan. 1990) (suit against pacemaker manufacturer; held: manufacturer's motion *in limine* to preclude discovery denied).

CHAPTER 12

Expert Evidence

CONTENTS

Expert Evidence

12.1 GENERALLY

12.1.1 Federal Rules of Evidence 702 and 703

In most toxic tort litigation, witness testimony, as to degree of risk and proximate cause, is crucial both to claimant's proof and to defendant's efforts to avoid liability. This testimony often takes the form of an opinion, and generally, only "expert" witnesses are permitted to give testimony in opinion form.

The Federal Rules of Evidence state this rule in these words: "[a] witness, who by his knowledge, skill, experience, training or education, has specialized knowledge that will assist the trier of fact in understanding the evidence or in determining a fact in issue may express an opinion about the matter."[1] Scientific evidence has been admitted on such diverse subjects as the effect of a pesticide upon the behavior of bees needed to pollinate plaintiff's crop[2] or whether a child's eye problems were attributable to stroke or administration of the DPT vaccination.[3]

Expert medical evidence has been admitted to prove causation issues ranging from the origination site of a worker's tumors pertinent to plaintiff's claim that the illness was related to asbestos exposure[4] to the potential for ventricular tachycardia or ventricular fibrillation from pacemaker spikes.[5] There is authority suggesting that characterization of plaintiff's injury as medical has not precluded introduction of testimony by experts in other disciplines. For example, testimony by a physicist has been held appropriate in plaintiff's suit claiming injury from an inadequate spinal shield during radiation therapy.[6]

The trial court's assessment of the admissibility of expert testimony requires the application of Federal Rules of Evidence 702 and 703 or their state law counterparts. Under Rule 702, a court may reject expert testimony on causation issues where it concludes that either (1) the witness is "unqualified to give expert testimony in the relevant field" or (2) the expert, irrespective of qualifications, has relied upon "an unreliable scientific technique".[7] Federal Rule of Evidence 703 requires that the basis an expert relies upon in forming an opinion be of "a type reasonably relied upon by experts in the particular field in forming opinions or inferences upon the subject."

The issues raised in challenges to expert testimony under the two rules may converge where the party opposing admissibility of an expert's testimony argues both that the "scientific technique" employed is unreliable and that the basis for the anticipated opinion is not of a type relied upon by others in that field of expertise. As one court acknowledged, in a given case it may be unclear whether "expert testimony depends on a reliable 'scientific technique', to be analyzed under Rule 702, or whether the testimony is based upon 'facts or data . . . of a type reasonably relied upon by experts in the particular field,' to be analyzed under Rule 703[.]"[8]

In deciding the sufficiency of the foundation of an expert's opinion, however, the court should not "substitute its views as to what constitutes reasonable reliance for those of experts in the field."[9] In addition, the factual assumptions upon which the expert relies in reaching her opinion may, in fact, be the subject of dispute between the parties, for most hypotheticals "are based on at least some disputed facts."[10]

The Rule 702 requirement of "helpfulness" to the factfinder "turns on whether the expert's 'technique or principle is sufficiently reliable so that it will aid the jury in reaching accurate results.'"[11] *In re Paoli R. R. Yard PCB Litigation,*[12] the court, while endorsing the rule that "helpfulness" imports "a quantum of reliability beyond that required to meet a standard of bare logical relevance,"[13] reversed the trial court's Rule 702 "reliability" rejection of plaintiff's expert's "meta-analysis". In so doing, the court suggested that in many instances only a hearing *in limine* (outside of the presence of the jury) on reliability issues will afford the trial court sufficient information to make a Rule 702 determination.[14]

Some courts suggest that they will be wary of receiving an expert's opinion as to a branch of medicine in which the physician has not specialized.[15] Nonetheless, in one suit in which plaintiff's representatives claimed a child's eye impairment was caused by administration of the DPT vaccine, a federal trial court permitted testimony of a defendant's expert whose training was in pediatric ophthalmology and not in either DPT or endotoxins.[16]

12.1.2 Sufficiency of Expert Testimony

A mere showing that the toxin is a possible cause of plaintiff's injury will not suffice to make out plaintiff's case of proximate cause. Even where the court has decided that an expert's testimony is admissible, the introduction of multiple expert opinions supporting plaintiff's proposition of causation may yet be found inadequate as a matter of law where the experts fail to conclude that there exists a reasonable probability that the toxin caused the injury.

Application of this general rule was evident in Rohrbough v. Wyeth Laboratories, Inc.,[17] the appeal of a DPT suit. At trial, one expert testified that he would not rule out the causal contribution of the vaccine; the second stated that the child's seizures could have resulted from the vaccination *or* from an ear infection and seizures; and the third established a *temporal* link between plaintiff's reaction and administration of the vaccination. Disregarding a later-filed affidavit of one expert seeking to bolster his conclusion, the court affirmed that plaintiffs "ha[d] not met their burden of coming forward with enough evidence that a jury could find that defendant's vaccine probably caused plaintiff's injuries.''[18]

12.2 VARIOUS FORMS OF EXPERT CAUSATION EVIDENCE

12.2.1 Generally

Admissible evidence of proximate cause may come from testimony of an examining physician[19] or from experts relying on medical evidence generated by others.[20] Absent other evidence of exposure to the toxin, the testimony of plaintiff's treating physician may also be dispositive as negativing plaintiff's proof of cause in fact. In one suit where plaintiff alleged that the drug cimetidine caused headache, nausea, and fatigue due to its property of inhibiting the passage of organophosphates and carbamates through the body, plaintiff's own physician negatived proof of causation by testifying that the interval between plaintiff's exposure to organophosphates and the onset of symptoms was too great for that exposure to be the cause.[21] Epidemiological evidence tending to support, or to refute, plaintiff's claim of proximate cause has been held admissible in suits claiming injury from products ranging from asbestos to intrauterine devices.[22]

Absent direct evidence of causation, plaintiff often relies upon epidemiological proof. In general, studies of adverse effects suffered by animal populations exposed to toxic substances are accorded lesser weight than is epidemiological or medical/diagnostic evidence.[23] Some courts have held that chemical structure analysis ''provide[s] little proof of causation.''[24]

12.2.2 Epidemiological Evidence of Causation

Epidemiology, a hybrid of science and medicine, ''uses studies to 'observe the effect of exposure to a single factor upon the incidence of disease in two otherwise identical populations.'''[25] While many courts have permitted nonmedical experts to testify as to epidemiological evidence of causation,[26] another authority suggests that plaintiff's exclusive resort to epidemiological proof may fail where plaintiff's expert is qualified to speak only of ''the incidence of disease over large population groups,'' and not whether defendant's product ''was the causative agent'' in the harm suffered by an individual plaintiff.[27]

Where the majority of epidemiological studies have tended to disprove plaintiff's claim that a particular product more probably than not caused a medical injury, some courts have held that an expert's opinion to the contrary, without another acceptable basis for such testimony, is excludable as having no adequate foundation.[28] Significantly, in other settings where vigorous disputation persists as to the teratogenicity of a pharmaceutical, there is authority that an expert's opinion that it is more probable than not that defendant's product caused injury has been held admissible even where most decisions on similar facts, and the majority of the expert community, have concluded otherwise. For example, in *DeLuca v. Merrill-Dow Pharmaceuticals, Inc.*,[29] a federal appellate court reversed a trial court's refusal to admit the testimony of plaintiff's expert that the morning sickness drug Bendectin was the proximate cause of her child's birth defects, even though the expert's opinion, and his interpretation of the epidemiological basis for it, was disputed by the weight of scientific opinion and was rejected by most courts considering like claims.[30]

There is agreement that the tort claimant need not establish the scientific standard of proximate cause, i.e., that there exists a 95% level of confidence "that the observed relationship is not related to random chance,"[31] i.e., proof of a "statistically significant" relationship between the two events.[32] In general, courts have held that plaintiff need not present epidemiological evidence conveying a statistically significant confidence level that defendant's product caused the injury.[33] Rather, a growing body of authority has concluded that plaintiff makes out a prima facie claim that defendant's product or process more probably than not caused the injury in question by introducing epidemiological evidence that a person so exposed is twice as likely to be afflicted with the disease or medical injury than would be a person not so exposed.[34]

Even where defendant's product or process produces concededly elevated risk factors in exposed populations, plaintiff relying upon epidemiological evidence often faces formidable barriers to proving that *his* injury was more likely than not cause by defendant. This dilemma was well described by Judge Weinstein in his 1984 opinion in *In re "Agent Orange" Product Liability Litigation,*[35] where he observes that "even if plaintiffs as a class could prove that they were injured by Agent Orange," regarding any individual claimant "[t]he probability of specific cause would necessarily be less than 50%[.]"[36]

For plaintiffs unable to demonstrate a risk elevation equal to or exceeding 2.0, the risk elevation showing that one exposed to the toxin is twice as likely to contract a particular disease than one not exposed, most courts still hold that this evidence, absent particularistic proof of individual causation, fails to establish proximate cause. Nevertheless, some decisions have held that the absence of a demonstrated risk factor of 2.0 or more is not a prerequisite of expert testimony on the issue of proximate cause where the studies themselves are not offered as evidence, but rather as a basis for the expert's opinion.[37]

12.2.3 *In Vitro, In Vivo,* and Other Studies

While epidemiological studies have gained a measured acceptance as bases for expert opinions, the Michigan federal trial court in *In re Bendectin Products Liability Litigation*[38] cautioned that scientific dispute upon exclusive reliance upon epidemiological evidence "necessitates . . . the conclusion that experts may reasonably rely upon other types of data when forming an opinion as to . . . teratogenicity[.]"[39]

Authority remains split as to the foundational value of nonepidemiological studies such as *in vivo, in vitro,* or structure activity analysis. In one Bendectin products liability action the federal trial court found plaintiff's proposal to prove the teratogenicity of Bendectin "through the presentation of . . . studies using *in vitro* testing, structure activity analysis, sales chart analysis, and animal studies" sufficient to deny defendant's motion for summary judgment.[40] In other suits, however, courts have found extrapolations from *in vivo* and *in vitro* animal studies to be inadequate as an expert's basis for an opinion on causation, as is "reanalysis" criticizing the methodology of other epidemiological studies.

12.3 BASIS FOR EXPERT OPINION

12.3.1 Generally

The trial court may remove an expert's testimony from jury consideration where it finds the basis for the opinion inadequate as a matter of law.[41] Such decisions have, for example, been upheld where a plaintiff's medical expert in an asbestos case had no knowledge of the degree of decedent's exposure.[42]

Toxic tort actions arising from expectant mothers' use of Bendectin have led to split authority as to what basis is sufficient for admitting medical expert testimony. In some suits, courts have found the basis insufficient for plaintiff's expert's testimony that Bendectin caused a child's birth defects.[43] In *DeLuca v. Merrill Dow Pharmaceuticals, Inc.*,[44] the appellate court adopted a liberal approach to contested theories of causation, and reversed a trial-court decision in which the court had refused the admissibility of plaintiff's expert's theory and data tending to show a link between ingestion of Bendectin and infant limb reduction. The appellate court held that an expert's opinion might be properly admissible even where it was contrary to the weight of scientific opinion where it was based upon reliable data of the type used by other experts in the field.[45]

12.3.2 Significance of Publication and Peer Review

Where an expert's testimony is based upon his or her own writing or other analysis, trial courts continue to ascribe weight to the presence or absence of prior publication of the findings.[46] Expert affidavits submitted on behalf of parties will be evaluated similarly for reasonable basis, and will be disregarded as expert evidence where unsupported by internal reference to or extrinsic support in any "medical literature, scientific publications, or personal experience" of the affiant.[47] An expert's incapacity to identify medical literature supporting a conclusion as to causation may represent more than an arguably inadequate basis under Federal Rule of Evidence 703, i.e., it may be interpreted as supporting the conclusion that the expert is not qualified.[48]

NOTES

1. *Celotex Corp. v. Tate,* 797 S.W.2d 197, Prod. Liab. Rep. (CCH) ¶ 12,673 at 37,998 (Tex. App. 1990) (in an action brought by the estate of a soundproofing worker who died of pleural mesothelioma, the appellate court left undisturbed the trial court's admission into evidence of a Dr. Castleman's testimony concerning "the knowledge of the dangers of exposure to asbestos that Celotex had or should have had at the time asbestos was being shipped to [decedent's] place of work[.]").
2. *Malensky v. Mobay Chemical Corp.,* 104 Or. App. 165, 799 P.2d 683 (Or. App. 1990) (entomologist's testimony on the repellant effect of defendant's insecticide Metasystox R (MSR) on bees essential to pollination of plaintiff's raspberry crop).
3. *Graham v. Wyeth Laboratories, Div. of American Home Products Corp.,* 906 F.2d 1399 (10th Cir. 1990).
4. E.g., *Haas v. Celotex Corp.,* [Not in F. Supp], 1990 WL 42835 (E.D. Pa. 1990) (plaintiff's experts testified that the tumor originated in the pleura, rather than the lung, thus indicating asbestos-related mesothelioma).
5. *Phillips v. Medtronic, Inc.,* [not in F. Supp], 1990 WL 58440 (D. Kan. 1990) (denying manufacturer's motion *in limine* to exclude plaintiff's opinion testimony).
6. E.g., *Ellingwood v. Stevens,* 564 So. 2d 932 (Ala. 1990).
7. *In re Paoli Railroad Yard PCB Litigation,* 916 F.2d 829, 855 (3d Cir. 1990). For Rule 702 purposes, a technique is considered "novel", as distinct from "unreliable", when its reliability is not so well established as to warrant recognition by judicial notice." *In re Paoli Railroad Yard PCB Litigation, id.* at 856 n.34.
8. *In re Paoli Railroad Yard PCB Litigation,* 916 F.2d 829, 856 (3d Cir. 1990) ("It can be difficult to determine whether the putative problem with scientific evidence lies in the underlying data itself or the method by which the data is analyzed."). There the court acknowledged that the distinctions between the inquiries each rule requires of the court are "of[t] times subtle if not strained." *Id.*).
9. *In re Bendectin Products Liability Litigation,* 732 F. Supp. 744, 749 (E.D. Mich. 1990) (declining defendant Bendectin manufacturer's argument that "plaintiffs' burden of proof on the causation issue can only be met through epidemiological evidence." *Id.* at 748.).
10. *Abbott v. Babcock & Wilcox,* 905 F.2d 201, 203, Prod. Liab. Rep. (CCH) ¶ 12,471 (8th Cir. 1990) (upholding admissibility of expert opinion as to causal connection between asbestos exposure and death by lung cancer despite witness's adoption of one calculation, among several, of the duration of the decedent's exposure).
11. *DeLuca v. Merrell Dow Pharmaceuticals, Inc.,* 911 F.2d 941, 957 (3d Cir. 1990), quoting 3 *Weinstein's Evidence* ¶ 702(03) at 702–735.
12. 916 F.2d 829 (3d Cir. 1990).
13. *In re Paoli, id.,* 916 F.2d at 857.

14. *In re Paoli, id.*, 916 F.2d at 858 ("We decline to define the exact level at which a district court can exclude a technique as insufficiently reliable." There is, however, "the requirement that there be a developed record and specific findings on reliability issues . . . [W]e cannot make such a determination on the record as it now stands. Not only was there no hearing, *in limine* or otherwise, at which the bases for the opinions of the contesting experts court be evaluated, but the experts were also not even deposed.").
15. E.g., *DeLuca v. Merrell Dow Pharmaceuticals,* 131 F.R.D. 71, (D.N.J. 1990) (finding inadequate basis for plaintiff's expert's opinion that other's use of Bendectin resulted in newborn's birth defects; court notes, among other deficiencies, that the expert relied upon epidemiological studies, but was not, himself, a specialist in epidemiology). The trial court's opinion was reversed by the Court of Appeals for the Third Circuit at 911 F.2d 941 (3d Cir. 1990).
16. See *Graham v. Wyeth Laboratories, Div. of American Home Products Corp.*, 906 F.2d 1399 (10th Cir. 1990) (admitting expert's opinion that visual impairment was due to stroke and not administration of defendant's vaccination).
17. 916 F.2d 970, Prod. Liab. Rep. (CCH) ¶ 12,629 (4th Cir. 1990).
18. *Id.*, ¶ 12,629 at 37,787.
19. Compare *Reeves v. Armstrong World Indus.*, 569 S.E.2d 1307 (Fla. App. 1990) (testimony by examining physician that plaintiff's asbestosis could conceivably have been caused by exposure to defendant's products was insufficient to support finding of proximate cause).
20. E.g. *Haas v. Celotex Corp.*, [not in F. Supp.], 1990 WL 42835 (E.D. Pa. 1990) (testimony of "distinguished array" of expert witnesses that decedent's tumor originated in the pleura, thus indicating its causal link to asbestos exposure).
21. *Latimer v. Smith Kline & French Laboratories,* 919 F.2d 301 (5th Cir. 1990) (court noting as well that outside of expert's report, there was no evidence of plaintiff's exposure to carbamates).
22. *Landrigen v. Celotex Corp.*, 579 A.2d 1268, Prod. Liab. Rep. (CCH) ¶ 12,652 (N.J. Sup. 1990) (asbestos); *Marder v. G.D. Searle & Co.*, 630 F. Supp. 1087 (D. Md. 1986) (intrauterine device); *In re Agent Orange Litigation,* 597 F. Supp. 740 (E.D.N.Y. 1984), aff'd, 818 F.2d 145 (2d Cir. 1987) (Agent Orange).
23. For example, in reversing a verdict for plaintiff claiming a skin pigmentation disorder caused by exposure to defendant's asphalt roofing tiles, the court in *Goode v. Tamko Asphalt Products, Inc.*, 783 So. 2d 184, Prod. Liab. Rep. (CCH) ¶ 12,318 (Tenn. 1989) notes that "[o]n cross examination, [plaintiff's expert] Mr. Green acknowledged that all but one of the articles or studies upon which he based his opinions involved experiments with rats, mice or guinea pigs." *Id.* at 36,305.
24. *Ambrosini v. Richardson-Merrell, Inc.*, _____ F. Supp. _____, Prod. Liab. Rep. (CCH) ¶ 12,189 (D.D.C. 1989). In that suit against defendant manufacturer of Bendectin, the court followed *Richardson v. Richardson-Merrell, Inc.*, 857 F.2d 823, 830 (D.C. Cir. 1988), where the appellate court considered the evidence of plaintiff's expert that the chemical structure of defendant's drug "revealed similarities between the antihistamine component of the drug and other antihistamines which have been found to cause birth defects in animals." Ambrosini, *id.* at 35,640.

25. *DeLuca v. Merrell Dow Pharmaceuticals, Inc.*, supra n. 19 (quoting Black & Lilienfeld, "Epidemiological Proof in Toxic Tort Litigation," 52 *Fordham Law Rev.* 732, 755 (1984). See generally Callahan, "Establishment of Causation in Toxic Tort Litigation," 23 Ariz. St. Law J. 604, 621–629 (1991).
26. *In re Paoli*, supra n. 20, at 862. Rejecting "rigid formalism" in applying Rule 702 standards for qualification of experts, the appeals court quoted the trial judge: "While it is true that an expert must demonstrate special competence to present expert testimony, there is no per se rule than nonphysicians are unqualified to testify about medical conditions of individuals exposed to chemicals."
27. *Id.*, ¶ 12,652 at 37,900.
28. *E.g.*, *Ealy v. Richardson-Merrell, Inc.*, 897 F.2d 1159, 1162 (D.C. Cir. 1990), a Bendectin case in which the court states that "[u]nder Rule 703, an opinion refuted by this scientific consensus [epidemiological studies] is inadmissible for lack of an inadequate foundation, in the absence of other probative evidence upon which to base this opinion.").
29. 911 F.2d 941, Prod. Liab. Rep. (CCH) ¶ 12,570 (3d Cir. 1990).
30. *Deluca, id.*, 911 F.2d at 953 ("[W]e do not view the absence of statistically significant findings or the great weight of contrary opinion as being relevant to the Rule 703 question posed here. Rule 703 is satisfied once there is a showing that an expert's testimony is based on the type of data a reasonable expert in the field would use in rendering an opinion on the subject at issue; it does not address the reliability or general acceptance of an expert's methodology.")
31. *Longmore v. Merrell-Dow Pharmaceuticals, Inc.*, 737 F. Supp. 1117 (D. Idaho 1990) (citing *Wells v. Ortho Pharmaceutical Corp.*, 788 F.2d 741, 745 (11th Cir. 1986) ("a distinction exists between legal sufficiency and scientific certainty.").
32. "Significance testing has a 'P value' focus; the P value 'indicates the probability, assuming the null hypothesis is true, that the observed data will depart from the absence of association to the extent that they actually do, or to a greater extent, by actual chance.' [citation omitted] If P is less than .05 (or 5%) a study's finding of a relationship supportive of the alternative hypothesis is considered statistically significant, if P is greater than 5% the relationship is rejected as insignificant. Accordingly, the results of a particular study are reported simply as 'significant' or 'not significant' or as P [less than] .05 or P [more than] .05." *DeLuca*, supra, 911 F.2d at 947.
33. *Longmore v. Merrell Dow Pharmaceuticals, Inc.*, 737 F. Supp. 1117, 1120 (D. Idaho 1990) (Plaintiff's epidemiological proof of causation "certainly does not require a confidence level of 95%, 90%, or even 80%. A cause and effect relationship may be deemed insignificant under stringent scientific standards, but nevertheless establish causation under legal standards.").
34. E.g., *In re Joint Eastern and Southern District Asbestos Litigation*, 758 F. Supp. 199, 202 (S.D.N.Y. 1991) ("For any risk factor, the epidemiologist . . . attempts to determine the magnitude of the risk: how much that factor will increase an individual's probability of contracting the disease. The magnitude is commonly expressed in terms of 'relative risk', the ratio of the number of occurrences of the disease in an exposed cohort to the number of occurrences

in an unexposed one. . . . Only when the risk level exceeds 2.0 can it be said that the one risk factor is more likely to cause the disease than any other factor affecting the unexposed cohort.'' See also *Cook v. United States,* 545 F. Supp. 306, 308 n.1 (N.D. Cal. 1982) (''[I]f the relative risk of GBS to nine-week vaccines is four, then 75% of all nine-week vaccines are vaccine linked. Once the relative risk rises above two, it becomes more probable than not that a given case was caused by the vaccine.'')

35. 597 F. Supp. 740 (E.D.N.Y. 1984).
36. *Id.*, 597 F. Supp. 740, 833–834. Based on the evidence before the court, Judge Weinstein explained: ''For example, plaintiffs as a class might be able to show that statistically, X% of the population not exposed to Agent Orange could have been expected to develop soft-tissue carcinoma, but that among those veterans who were exposed to Agent Orange X + Y% suffer from soft-tissue carcinoma. If Y is equal to or less than X and there is no meaningful 'particularistic' or anecdotal proof as to the vast majority of plaintiffs, virtually no plaintiff would be able to show by a preponderance of the evidence that his or her cancer is attributable to Agent Orange rather than being part of the 'background' level of cancer in the population as a whole. The probability of specific cause would necessarily be less than 5% based upon the evidence submitted.'' *Id.*
37. *Grassis v. Johns-Manville Corp.*, 591 A.2d 671 (N.J. App. Div. 1991) (testimony of physician-epidemiologist concerning relationship between asbestos exposure and colon cancer).
38. 732 F. Supp. 744 (E.D. Mich. 1990).
39. *Id.*, 732 F. Supp. at 749.
40. *In re Bendectin Products Liability Litigation*, supra n. 38 at 748.
41. E.g. *Landrigan v. Celotex Corp.*, 243 N.J. Super. 449, 579 A.2d 1268 (N.J. Super. App. Div. 1990) (expert medical witness with no knowledge of the degree of decedent's exposure to defendant's asbestos products).
42. *Landrigan v. Celotex Corp.*, 243 N.J. Super. 449, 579 A.2d 1268 (N.J. Super. App. Div. 1990).
43. E.g., *DeLuca v. Merrell Dow Pharmaceuticals,* 131 F.R.D. 71, (D.N.J. 1990), reversed 911 F.2d 941, Prod. Liab. Rep. (CCH) ¶ 12,570 (3d Cir. 1990) (court noting that reports on which plaintiff's expert relied were either irrelevant or reached conclusions differing from those reached by witnesses).
44. 911 F.2d 941, Prod. Liab. Rep. 12,570 (3d Cir. 1990).
45. *Deluca v. Merell Dow Pharmaceuticals, Inc.*, 911 F.2d 941, Prod. Liab. Rep. (CCH) ¶ 12,570 (3d Cir. 1990) at 37,520 (''In *Downing*, we explicitly rejected reliance upon the 'general acceptance' test of admissibility, most prominently articulated in *Frye v. United States,* 293 Fed. 1013 (D.C. Cir. 1923). We did so, among other reasons, because the general acceptance test was too vague and malleable to yield consistent results, and because its nose-counting emphasis often led to the exclusion of helpful evidence in contradiction to the spirit of the Federal Rules of Evidence 753 F.2d at 1236–1237. Thus, under *Downing,* Dr. Done's opinion cannot be excluded simply because the weight of scientific opinion leans against him. At the same time, the degree to which contrary opinion dominates the relevant literature is not wholly irrelevant to the reliability inquiry mandated by *Downing. Id.* at 1238.'').

46. E.g., *Richardson v. Merrell Dow Pharmaceuticals, Inc.*, 857 F.2d 823, 831 (D.C. Cir. 1988), *cert. denied,* ____ U.S. ____, 110 S. Ct. 218 (1989). In the appeal of that action claiming the drug Bendectin caused the minor-plaintiff's limb reduction defects, the court affirmed the trial court's conclusion that plaintiff's expert's conclusion was based upon an inadequate foundation, and noted: "Only by recalculating the data was Dr. Done able to obtain what he considers a statistically significant result. Moreover, the studies rejected by Dr. Done had been published in peer-reviewed scientific journals, while Dr. Done has neither published his recalculations nor offered them for peer review." (footnotes omitted).
47. *Maddy v. Vulcan Materials, Co.*, 737 F. Supp. 1528, 1533 (D. Kan. 1990).
48. E.g., *Maddy v. Vulcan Materials Co.*, 737 F. Supp. 1528, 1533 (D. Kan. 1990) ("Dr. Pollard repeatedly stated during his deposition that he has no expertise in the area. He was unable to identify any medical literature to support the conclusion that chemicals allegedly emitted by Vulcan had caused or aggravated Maddy's respiratory problem, nor was he able to point to any personal experience that would support such a conclusion.").

CHAPTER 13

Citizen Suits

CONTENTS

Citizen Suits

13.1 GENERALLY

Actions to enjoin environmental harm and enforce civil penalties, and sometimes to recover remedial costs, commonly called "citizen suits", may be permitted under certain federal[1] or state statutes.[2] Some of the earliest statutory provisions for citizen suit private cost recovery were codified in the Clean Air Act Amendments (CAA) of 1970.[3] Only in the past decade, however, has there been widespread resort to citizen suit relief.[4] The principal impetus for citizen suit enforcement of environmental statutes was the perception that citizen suits would enhance and supplement federal enforcement[5] and that individuals pursuing citizen suits should be considered "welcome participants in the vindication of environmental interests."[6] While Congress has attempted to encourage citizen suits by slackening traditional jurisdictional requirements,[7] such as eliminating a minimum amount in controversy requirement and disregarding citizenship, there remains the countervailing reluctance to burden the courts with such suits. In partial response to the latter consideration, the citizen suit claimant may not seek individual compensatory damages.[8]

As most statutes containing citizen suit provisions do not specifically spell out the purposes of the remedies or the mechanisms for their enforcement, two concerns persist. The first concern is judicial lawmaking. The lack of guidance in the statutes regarding enforcement could force courts to determine the content of the regulatory requirements, resulting in potentially inconsistent enforcement and loss of administrative control.[9] The second concern is that availability of citizens suits may make attainment of the optimal mix of cooperation and deterrence under environmental laws difficult, insofar as inviting private litigants to enter the arena of enforcement of environmental statutes presupposes private plaintiff dissatisfaction with the vigor and effectiveness of agency enforcement of the same statutes.[10]

13.2 PREREQUISITES TO A CITIZEN SUIT

13.2.1 Standing

The citizen suit complainant must demonstrate "standing" to bring the action. Standing in citizen suits has been liberally construed due to the general terms employed by the statutes. Generally any citizen who has an interest which is adversely affected by violation of a statute including a citizen suit provision may commence a citizen suit on his own behalf. For example, the Surface Mining Control and Reclamation Act[11] provides that "any person having an interest which is or may be adversely affected may commence a civil action on his own behalf to compel compliance with this Act."[12]

However, to have standing there must be more than a "mere interest;" a citizen must allege an injury in fact, namely, that the party seeking redress must be injured himself. In *Sierra Club v. Morton,*[13] the Supreme Court denied standing to the environmental group due to the complainant club's failure to allege injury to itself. Although the *Sierra Club* court held that the environmental group was not a proper plaintiff in that particular case, the Court did not, however, find that groups or organizations as entities could not have standing to bring a citizen suit. The court stated that organization may have standing to bring a citizen suit so long as there is actual injury to, or adverse effect upon, members.[14]

Applying the *Sierra Club* standard, in *Public Research Interest Group of New Jersey v. Yates Ind.,*[15] a federal trial court found that an environmental interest group had standing where some of its members lived downstream from a pollutant discharge point. The defendant claimed that there was no standing as the persons actually affected were not parties to the suit. Although the actual members whose interest were being adversely affected were not parties to the suit, the court held that when the standing of a group is challenged the plaintiff "simply ha(s) to establish that they represented the interest of the members who had standing."[16]

Most citizen suit provisions also allow citizens to intervene as a matter of right in civil suits filed by the government in federal court.[17] The Administrator, if not a party, may also intervene as a matter of right.[18]

13.2.2 Jurisdiction

Citizen suit provisions provided for by federal statute confer federal question jurisdiction on actions where citizens allege violations of the statute. Such jurisdiction is granted under most citizen suit provisions without regard to the amount in controversy or the citizenship of the parties.[19] In response to the lenient jurisdictional requirements, some federal courts of appeal have held that when the procedural requirements such as notice are not met there is no jurisdiction.[20] Other federal courts have been less strict.[21]

One key element in determining jurisdiction is timing, as most statutes are worded to include the phrase "against any person alleged to be in violation."[22] The Supreme Court addressed this language in *Gwaltney of Smithfield, Ltd. v. Chesapeake Bay Foundation.*[23] In *Gwaltney,* the last alleged violation occurred in 1983, and the complaint was filed in early 1984. The Court held that jurisdiction did not exist because the suit was based upon a wholly past violation. The Court stated that the language "alleged to be a violation" means that there must be an ongoing violation at the time the complaint was filed. In the absence of an ongoing violation, the Court concluded, the petitioner's request for an injunction was moot. Only the Administrator might bring an action for civil penalties involving wholly past violations. A citizen suit plaintiff could not seek civil penalties for wholly past violations because there is no jurisdiction.

The Supreme Court did, however, conclude that intermittent or continuous violations could be considered "ongoing". Reasoning that intermittent violations carried with them the possibility of recurrence, the court observed that in such circumstances, a petitioner could file a citizen's suit claim, even where at the moment of filing there was not an actual violation.

In *Atlantic States Legal Foundation v. Tyson Foods, Inc.*,[24] the federal appellate court interpreted *Gwaltney* in the situation where there is a violation at the time the complaint is filed, but the violation has, thereafter, been permanently corrected. In *Atlantic,* Tyson Foods, Inc. was in violation of the Clean Water Act (CWA) from May 1986 through February 1988. Atlantic filed its complaint August 7, 1987. Tyson argued that according to *Gwaltney,* once it came into compliance with the Act an injunction was moot, and therefore the violation was wholly past, rendering petitioner's claim without subject-matter jurisdiction. The court of appeals, disagreed, pointing out that *Gwaltney* dealt with the presence or absence of jurisdiction in the first instance, not mootness. It continued by explaining that *Gwaltney* did not apply to the matter before it, reasoning that "alleged to be in violation" and "ongoing violation" attach to the time of complaint, not the time of trial. If the violation was ongoing when the complaint was filed, the court held, that it would have subject matter jurisdiction over petitioner's citizen's suit claim. The appeals court explained that the subsequent end of the violation does make an injunction moot, but it does not deprive the court of jurisdiction to consider plaintiff's claims for civil penalties for the past violations.

The key distinction between *Gwaltney* and *Atlantic* is the timing of the complaint. If the complaint in *Atlantic* was filed after the violation ceased to exist, then there would have been no jurisdiction in the court for plaintiff's pursuit of civil penalties.

The CAA Amendments of 1990 contains the first citizen suit provision which may allow a plaintiff to proceed against a violator for wholly past violations. The language of the amendment reads "against any person . . . who is alleged *to have* violated (if there is evidence that the alleged violation has been repeated) or to be in violation."[25] It is interesting to note that the legislature included the parenthetical because in the past, a repeated violation was evidence of an intermittent violation that would likely occur again in the future even if it was not currently going on. As such it could be considered a present violation. On one hand it is possible to read the new language as being consistent with the past interpretation of the statute. On the other hand, this is the first time the legislature explicitly used the past tense i.e., to have violated. The use of the past tense could indicate that an action could be maintained for repeated past violations even if the defendant could prove that the problem has since been permanently corrected.

13.3 PROCEDURAL REQUIREMENTS

13.3.1 Notice

Before a citizen suit may be filed, notice must be given 60 days in advance to the alleged violator and the appropriate governmental authority. For instance, the CAA prohibits actions from being commenced prior to 60 days after notice has been served to: (a) the Administrator; (b) the state in which the violation occurred; and (c) any alleged violator.[26] The purpose of this notice period is to give the Environmental Protection Agency and the alleged violator time to take corrective measures. However, where there is an imminent danger under some statutes notice is not required.[27]

As traditional barriers to jurisdiction, such as amount in controversy and citizenship, are not included in citizen-suit provisions, defendants in a citizen suit claim will often vigorously attack the petitioner's satisfaction of procedural requirements. One decision addressing the relationship between notice and jurisdiction is *Martin v. Kansas Board of Regents.*[28] In *Martin*, the Martin family lived on property immediately adjacent to the University of Kansas. On university premises was a landfill into which the university allegedly dumped radioactive waste, contaminating the groundwater and the well water upon which the Martin residence relied. The Martins filed claims under the CWA, Comprehensive Environmental Response, Compensation, and Liability Act (CERCLA), and Resource Conservation and Recovery Act (RCRA). They filed notices of those claims on July 16, 1989, and then filed their complaint 18 days later. The Board argued that the Martins did not give 60-day notice as required by statute and that the court, therefore, had no jurisdiction over their claim. The federal trial court concluded otherwise. In regard to the CWA and CERCLA, the court found it had jurisdiction, reasoning that Congress did not intend that any failure to conform to the 60-day notice requirement should constitute an absolute bar. Dismissing the claim, and requiring the Martins to file later, would be "excessively formalistic", the court stated and cited numerous decisions supporting its conclusion that remedial statutes intended to serve the public health and welfare should not be construed rigidly.

The court further noted that in regard to RCRA, although there is a general 60-day notice requirement, the statute provides for an exception. When hazardous waste presents an "imminent and substantial endangerment to health or environment,"[29] a complaint may be filed immediately after notification.

13.3.2 Lack of Diligent Prosecution

An additional prerequisite to private prosecution of a statutory claim under a citizen suit provision is that at the time of filing the complaint there is no government authority "diligently prosecuting" the alleged violation in a court of the U.S.[30] Courts have differed on the interpretation of "diligently prosecuting" and "courts".

In deciding when an action is being diligently prosecuted, courts look to the particular facts of the action. Included as being a "diligent prosecution" is a state action under a state environmental law comparable to federal statute. A federal trial court addressed the question of what represents a comparable state environmental law in *North and South Rivers Watershed Assoc. v. Scituate.*[31] In that suit, plaintiff, a nonprofit membership organization, brought a citizen suit against the town of Scituate alleging violations of the federal CWA arising from sewage overflow into an estuary. The town was granted summary judgment from the district court on the grounds that the state had commenced and was vigorously prosecuting the violation under a comparable state environmental statute, inasmuch as the Massachusetts Department of Environmental Protection had issued the town an administrative order pursuant to the Massachusetts CWA.

Plaintiffs argued that the order promulgated under the state statute was not comparable to orders available in enforcement of the federal CWA, as the state law, unlike the federal, did not provide for civil penalties. The federal court disagreed, pointing out that although the one section of the state statute did not authorize civil penalties, another provision did. The court continued by stating that to find commensurate state authority the court must look at the entire statute within the context of its overall enforcement scheme. The court further found that "comparable" does not mean that the federal administrative agency has to confer power upon the state agency or otherwise "approve" the state law. What comparable does mean, the court concluded, was that the state law must include public notice, participation, penalty assessment, and judicial review.

The court also provided guidelines for interpretation of what constitutes "diligent prosecution". It suggested that simply because a comparable statute exists, and an order has issued, it does not necessarily follow that the prosecution is diligent. A diligent prosecution, the plaintiff argued, requires an order that demands compliance with the federal act by a specific date and provides civil penalties for violations. The court did not take such a restricted view, and while considering the factors the plaintiff suggested, nonetheless found that an order is acceptable if it "represents a substantial, considered and ongoing response to the violations."[32] The *North and South Rivers Watershed Assoc.* court observed further that as citizens suits are intended as a supplement to, and not a substitute for, government enforcement, a certain amount of deference should be given to state statutes and orders because the states have the primary responsibility to remediate pollution.

In evaluating what constitutes "diligent prosecution", courts have also been required to determine what constitutes a "court". In *Baughman v. Bradford Coal Co.*, the Court of Appeals for the Third Circuit stated that an administrative board can be a court when it has court-like procedures and the power to afford relief equivalent to that of a court.[33] While the *Baughman* court stated these criteria, it found the administrative body it was dealing with did not meet them, for the administrative body could only levy fines equal to one half of what a federal court could order, and could not enjoin violations. Another federal appellate court has interpreted the meaning of "court" more strictly, reasoning that the language was unambiguous and holding that administrative enforcement actions are court actions.[34] Other federal courts of appeal have abided by this strict interpretation.[35]

NOTES

1. *Public Interest Research Group of New Jersey v. Yates Industries, Inc.*, 757 F. Supp. 438, 33 E.R.C. (BNA) 1142 (D.N.J. 1991) (environmental group's suit under federal Clean Water Act arising from discharge of pollutants into waterway).
2. See, e.g., *General Electric Environmental Services, Inc. v. Envirotech Corp.*, 763 F. Supp. 112, 33 E.R.C. (BNA) 1212 (M.D. Pa. 1991) (Pennsylvania property owner's claim under that state's Hazardous Sites Cleanup Act to recover PCB cleanup costs from prior property owner).
3. 42 U.S.C. §§ 1587-1588a (1970).
4. Comment, "The Rise of Citizen Suit Enforcement of Environmental Law: Private and Public Attorneys General," 81 N.W.U.L. Rev. 220, 221–222 (1987).
5. *Friends of the Earth v. Carey*, 535 F.2d 165, 172 (2d Cir. 1976) [quoting S. Rep. No. 1196, 91st Cong., 2d Sess. 35–36 (1970)].
6. *Friends of the Earth v. Carey, id.*, 535 F.2d at 172.
7. See Clean Air Act, 42 U.S.C. § 7604(a) (1990).
8. For example, pursuant to 42 U.S.C. § 7604(g) (1990), penalties received shall be deposited in the U.S. Treasury for licensing and other services. Also, the jurisdictional clause, § 7604(a), provides district courts with the power to enforce the act's injunctive orders and civil penalties.
9. Comment, "The Rise of Citizen Suit Enforcement of Environmental Law: Private and Public Attorneys General," 81 N.W.U.L. Rev. 220, 223 (1987).
10. *Id.*
11. 30 U.S.C. § 1270.
12. 30 U.S.C. § 1270(a).
13. 405 U.S. 727 (1972).
14. *Sierra Club v. Morton*, supra, 405 U.S. 727 (1972). See also *Sierra Club v. SCM Corp.*, 747 F.2d 99 (2d Cir. 1984).
15. 757 F.Supp 438 (D.N.J. 1991).
16. *Id.*, 757 F. Supp. at 443.
17. E.g., 42 U.S.C. § 7604(b)(1)(B) (Clean Air Act).
18. 42 U.S.C. § 7604(c) (1990).
19. E.g., Toxic Substances Control Act, 15 U.S.C. § 2619(a) ("subject matter jurisdiction regardless of the amount in controversy or the citizenship of the parties").

20. Comment, "The Rise of Citizen Suit Enforcement of Environmental Law: Private and Public Attorneys General," 81 N.W.U.L. Rev. 220, 334 n. 33 (citing authority from First, Sixth, Seventh and Ninth Circuit Courts of Appeal).
21. Id., 81 N.W.U.L. Rev. at 335 (Second and Third Circuit Courts of Appeal).
22. E.g., Clean Water Act, 33 U.S.C. § 1365(a).
23. 484 U.S. 49 (1987).
24. 897 F.2d 1128, 31 E.R.C. (BNA) 1201 (11th Cir. 1990).
25. 42 U.S.C. § 7604(a) (1990).
26. 42 U.S.C. § 7604(b)(1)(A).
27. Surface Mining Control and Reclamation Act, 30 U.S.C. § 1270(b)(2); Clean Water Act, 33 U.S.C. § 1365(b).
28. [Not in F. Supp.], 1991 WL 33602, 32 E.R.C. (BNA) 1944 (D. Kan. 1991).
29. 42 U.S.C. § 6973(a) (1984). This language describes the authority of the Administrator. Another section, which applies to citizen plaintiffs, allows for suit immediately after notice is given if the violation involves 42 U.S.C. § 621 et seq. (hazardous waste). 42 U.S.C. § 6972(b).
30. This requirement usually follows the section on notice and is listed under a heading such as "Limitations on Actions." See Clean Air Act, 42 U.S.C. § 7604(b)(1)(B). Most statutes require that the prosecution be a civil one, although a few statutes also explicitly include criminal prosecutions under this section. See, e.g., Deepwater Ports Act, 33 U.S.C. § 1365(b)(1)(B).
31. 755 F. Supp. 484 (D. Mass. 1991).
32. *Id.*, 755 F. Supp. at 487.
33. 592 F.2d 215, 219 (3d Cir. 1979). See also *Sprig v. Fritzsche, Dodge & Olcott Inc.*, 22 E.R.C. (BNA) 1721 (3d Cir. 1985).
34. *Friends of the Earth v. Consolidated Rail Corp.*, 768 F.2d 57, 59, 62 (2d Cir. 1985) (interpreting the federal CWA). See also *Vernon Village, Inc. v. Van Ostrand, High Manor, Inc.*, 755 F. Supp. 1142 , 1152 (D. Conn. 1990).
35. *Sierra Club v. Chevron U.S.A., Inc.*, 834 F.2d 1517, 1519 (9th Cir. 1987). Another procedural barrier contained in some statutes is the provision that when the EPA Administrator has assessed an administrative penalty a citizen suit may not be filed. E.g., 33 U.S.C. § 1319(g)(6)(B).

CHAPTER 14

Defenses

CONTENTS

Defenses

14.1 REBUTTAL AND AFFIRMATIVE DEFENSES — GENERALLY

The defendant manufacturer or seller of a toxic product will raise two tiers of rebuttal to a claimant's cause of action. The first tier will be defendant's refutation, where possible, of each aspect of plaintiff's prima facie case. The earlier chapters discussing claims in negligence, warranty, strict products liability, liability for abnormally dangerous activities, and nuisance or trespass, describe the elements of each of these claims. A defendant's general, or rebuttal, defense, is in effect, a denial of each element of plaintiff's claim for which there exists a good faith basis to deny it.

The second tier of defendant's defenses are the so-called "affirmative" defenses. They are denominated "affirmative" because defendant must raise them affirmatively and specifically, by either pretrial motion or at trial. Be the matter tried to a jury or to a single judge, the factfinder cannot find for defendant on the basis of an affirmative defense, even where there is sufficient evidence to permit such a finding, unless defendant raises the defense explicitly by pleading or motion.

14.1.1 Rebuttal By Categories of Claim

Choosing plaintiff's negligence claim as an example, defendants might take the following approach to its general, first-tier defenses. Plaintiff in a negligence claim must produce positive proof of defendant's duty to plaintiff, breach of that duty, the proximate causal contribution of defendant's act to the injurious incident, and plaintiff's damages resulting therefrom. Depending upon the facts in any particular claim, defendant's general, rebuttal defense may commence by introduction of evidence that: (1) defendant owed no duty to plaintiff; (2) if there was a duty, defendant did not breach that duty; (3) if there was duty and breach, the breach was not the proximate cause of plaintiff's injury; and (4) negligence law does not provide damages for injuries of the nature plaintiff suffered.

For the variety of legal theories often named in successive counts of the typical toxic torts complaint, the defendant styles the general, first-tier defense similarly. For example, if plaintiff claims that effluent from defendant's manufacturing process caused damages in trespass to plaintiff's property, defendant first argues, where supportable, (1) that it did not intend the act; (2) that the act did not cause an invasion or interference with plaintiff's property rights; (3) that plaintiff's property right is not of a nature that qualifies plaintiff to receive damages in trespass; and so on.

14.1.2 Affirmative Defenses — Generally

Defendant will also invoke, where supported by the facts of the case, such affirmative defenses as contributory negligence, assumption of the risk, or misuse of the toxic product. Under the law of many states, such affirmative defenses must be pleaded specifically even though conceptually their proof pertains as well to defendant's first-tier defenses. For example, defendant's proof of product misuse is applicable to the affirmative defense by that name, but it pertains as well to defendant's refutation of plaintiff's prima facie claim that the toxic product was the proximate cause of plaintiff's harm or loss.

14.1.2.1 *Contributory Negligence*

To show contributory negligence, defendant must demonstrate that the injured party acted without due care for his personal safety, i.e., that he was *negligent* in protecting himself.[1] It would, in some circumstances, suffice to show that plaintiff was not aware of the risks, while a reasonably alert person would have been.

An injured party's failure, or his employee's, to read and heed a warning or instructions for safe use of a product may constitute contributory negligence sufficient to bar the claim. For example, in *Champs Convenience Stores, Inc. v. United Chemical Co.*,[2] the store sued a cleaning product seller that mistakenly delivered automotive cleaner, and not a floor cleaner, to the premises. Use of the wrong product upon the floors caused damage, but defendant avoided liability by proving that the store's employee was contributorily negligent in applying the wrong product. The evidence at trial showed that the employee (1) failed to read the product's label before beginning to mop the floor with it and (2) did not pause to check the label, the product name, or the instructions for use even after noticing a bad odor from the product. On such proof, the appeals court held, the employee's inattention was a proximate cause of the property damage suffered and barred the store's recovery.

In some circumstances, the means by which a toxic product was used by exposed employees precludes a finding that they acted without due care for their personal safety. In *Eagle Picher Industries v. Balbos*,[3] the court held that there was no evidence upon which the trial court could instruct the jury that it could have found decedent, a shipyard erector working in proximity to asbestos products, contributorily negligent. Defendant argued that the jury should have been permitted to find that decedent failed to see, read, and heed the warnings defendant placed upon its products starting in 1964. Rebuffing defendant's argument, the court observed that in his work as an erector, decedent would never have observed defendant's products in their original packaging. Rather, the only way decedent would have learned of the risks associated with asbestos exposure would be "when he saw those who worked directly with the products don respirators and other protective equipment."

As a consequence, the appeals court stated, "the worker could not have been contributorily negligent for having 'fail[ed] to see, read and heed' the warnings."

The viability of contributory negligence as a complete defense exists more broadly in suits claiming toxic tort damage to property alone.[4]

14.1.2.2 *Assumption of the Risk and Product Misuse*

To prove assumption of the risk, defendant must establish that the plaintiff was personally aware of the specific risks involved in employing the product or the process, and that he nonetheless knowingly and voluntarily proceeded to expose himself to the hazard.[5] Originally, the distinction between contributory negligence and assumption of the risk had a greater significance than it does today. As one court summarized, "[w]hile contributory negligence was never a defense in strict product liability cases, misuse of the product and assumption of the risk were defenses which could be interposed as an absolute bar to recovery."[6] Under prevailing modern doctrines of comparative fault, all but the most aggravated forms of product misuse or assumption of the risk do not altogether bar plaintiff's claim, but rather operate to reduce plaintiff's award by the proportion plaintiff's substandard conduct contributed to his injury or loss.

The key inquiry in the seller's defense that the injured party assumed the risk is the injured party's *actual, subjective* appreciation of the hazard. Thus, for example, a worker's conscious disregard of manufacturer warnings about exposure to its products may constitute a misuse absolving the seller of a claim in failure to warn adequately.[7]

At the trial of a toxic tort case, examination of these factors is quite meticulous. For example, in *Byrne v. SCM Corp.*,[8] the appeals court agreed that the trial judge had properly disallowed the manufacturer's defense that plaintiff, a painter overcome by fumes from epoxy paint, had assumed the risk. The appeals court summarized the reasons in this language: "The evidence does not establish the plaintiff appreciated, or could appreciate, the risk involved. Plaintiff may have once complained that the [protective] mask was weather-cracked . . . [and] not working. . . . [Even so,] no one appreciated that the mask being used was inappropriate. Nor is there evidence that plaintiff understood what defendants meant by 'Good Ventilation[.]' . . . Plaintiff was not advised of the need to utilize an air-supplied respirator, and therefore, he could not possibly have proceeded in the face of a known danger."

Defendant may employ the related defense of product misuse by showing that plaintiff's harm resulted from a use of the product for a purpose, or by means, not intended by the manufacturer, and where the manufacturer could not reasonably have foreseen such a misuse. The facts of a particular case will often create the potential for defendant to argue both assumption of the risk and misuse. In other circumstances, misuse alone may be fairly argued, such as, for example, plaintiff misused the product but did not subjectively appreciate the nature or severity of the risk.

14.2 COMPARATIVE FAULT STATUTES

The law of "comparative fault" varies from state to state. In most instances, where plaintiff's deficient conduct *and* an unreasonably dangerous condition in the product *both* operated to cause the harm, plaintiff's conduct will be evaluated to reduce his damage recovery in the proportion the jury finds plaintiff may be fairly held responsible.

Today, in states applying principles of "pure" comparative fault, plaintiff's contributory negligence, assumption of the risk, or misuse of the product will not act to totally bar plaintiff's claim unless plaintiff's conduct is seen as the sole producing cause of the harm. In such states, "pure" comparative fault permits plaintiff to recover money damages even where plaintiff's comparative causal contribution to his own injury is greater than the proportionate contribution of defendant.

In other states, plaintiff's contributory negligence, assumption of the risk, or product misuse may bar his claim altogether only where the jury finds that plaintiff's misconduct, in terms of its contribution to the eventual harm, was as great as or greater than defendants. In these states, application of a "modified" comparative fault scheme means that plaintiff's conduct may preclude recovery only where it is seen as an equivalent cause of the harm that was the condition of defendant's product or process. The law of yet another group of states operates to bar plaintiff's claim only upon defendant's showing that plaintiff's substandard conduct, in terms of its contribution to the injury, was more responsible for the injury than was the conduct of defendant.

14.3 LIMITED IMMUNITY OF GOVERNMENT CONTRACTORS

In *Boyle v. United Technologies Corp.*,[9] the Supreme Court defined a limited immunity from design defect claims for sellers of products to the government where: "(1) the United States approved reasonably precise specifications; (2) the equipment conformed to those specifications; and (3) the supplier warned the United States about the dangers in the use of the equipment that were known to the supplier but not to the United States."[10]

In accord with various statements of the government contractor defense prior to *Boyle*, decisions following it reaffirm that "equipment" may mean virtually any product sold to the government pursuant to its specification.[11] That a manufacturer produces a potentially toxic substance for both government and nongovernment use alike should not affect applicability of the government contractor defense where injury or illness allegedly arises from a governmental use of the product.[12] On the other hand, the manufacturer who, pursuant to specification, provides a product to the government, and also sells the identical product to the private sector, should not be permitted to invoke the government contractor defense in a claim brought by a person injured by use of or exposure to the privately sold product.

The reach of the government contractor defense extends beyond design defect claims to include allegations of failure to warn adequately of a product's risks.[13] In the words of one court, "stripped to its essentials, the military contractor's defense under *Boyle* is to claim, 'The Government made me do it.' *Boyle* displaces states law only when the government, making discretionary, safety-related military procurement decisions contrary to the requirements of state law, incorporates this decision into a military contractor's contractual obligations, thereby limiting the contractor's ability to accommodate safety in a different fashion."[14]

The court in *Jackson v. Deft, Inc.*[15] also applied the *Boyle* criteria in a toxic tort context. In that suit, the appeal of a claim brought by a worker who claimed to have developed asthma and other disabilities as a result of his exposure to fumes from paint manufactured by defendants according to government specifications. There the appellate court reversed the trial court's grant of summary judgment to defendants. The trial court had stated that the manufacturer was immune from liability due to the government contractor defense and that its warnings on the paint containers were adequate as a matter of law. Regarding the limited immunity afforded government contractors, the appellate court held that although evidence demonstrated that "certain warnings were required by the military specifications," the evidence did not establish "that the specifications placed any limitation on additional information from the manufacturers to users of their products."[16]

In denying government contractor claims for indemnification by the government, some courts have stated that to permit indemnity claims by, for example, asbestos contractors seeking to relieve themselves of a portion of the personal injury awards paid out to shipyard workers and other employees, would "reduce the government's immunity and . . . appea[r] logically inconsistent with [*Boyle*]." As the court in *Lopez v. A.C.& S., Inc.*[17] stated, if a government contractor is truly exposed to claims "by reason of conforming to government specifications and, if the otherwise carefully prescribed conditions are met, there should be no damage awards on which to base claims for government indemnity or contribution."

14.4 PREEMPTION — GENERALLY

Congress may, by statute, regulate an area of potential toxic tort litigation so as to suggest that no inconsistent state remedies, common law or statutory, should be permitted. Upon occasion, as in the 1976 Medical Device Amendments, the federal statute *expressly* preempts inconsistent state regulation or litigation. In a variety of other circumstances, the federal statute does not by its terms explicitly preclude state remedies, or while arguably precluding some litigation under state law, leaves open to question what, if any, state remedies may be litigated.

Statutes of the latter nature raise issues of *implied* preemption, and have required courts to evaluate congressional intent, purpose, policy, and evidence, in the legislative record, of whether congress intended that its statute constitute the exclusive rule of law governing certain potentially toxic products or processes. A recent and representative example of evaluation of implied preemption is application of the Federal Insecticide, Fungicide, and Rodenticide Act (FIFRA) to claims alleging toxic harm from pesticide exposure.

14.4.1 Express Preemption

14.4.1.1 *The 1976 Medical Device Amendments*

In the 1976 Medical Device Amendments to the Food, Drug, and Cosmetic Act, Congress vested the Food and Drug Administration (FDA) with exclusive authority over medical device design, performance, and warnings or instructions for use.[18] Claims alleging tampon manufacturer failure to adequately warn users of potential health risks and instruct them in means of avoiding those risks have been held to be preempted by the medical device statute and regulations thereunder,[19] a construction of FDA regulations that the Supreme Court left undisturbed in *Berger v. Personal Products, Inc.*[20]

14.4.2 Implied Preemption

14.4.2.1 *Federal Insecticide, Fungicide, and Rodenticide Act*

FIFRA states that a "state shall not impose or continue in effect any requirements for labeling or packing in addition to or different from those required under this subchapter."[21] Notwithstanding this clear statement of congressional intent that FIFRA operate as the principal guide of manufacturer obligations regarding the labeling of pesticides and related products, in *Wisconsin Public Intervenor v. Mortier*,[22] the Supreme Court held that FIFRA did not preempt local regulation of pesticide use.

Wisconsin Public Intervenor is significant in its removal of a hindrance to city or town control or even ban of pesticides. Similarly, in *Riden v. ICI Americas, Inc.*,[23] the federal trial court held that FIFRA neither expressly nor impliedly preempts state law failure to warn claims against manufacturers of pesticides. Prior to *Wisconsin Public Intervenor* other courts had reached the opposite conclusion, finding that the legislative history of FIFRA demonstrated that Congress impliedly preempted local regulation of pesticides.[24]

14.4.2.2 *National Manufactured Housing Construction and Safety Standards Act of 1974*

The National Manufactured Housing Construction and Safety Standards Act (NMHCSSA)[25] and the regulations thereunder provide emission, exposure, labeling, and warning requirements upon the distribution of a variety of materials used in home construction, including mobile home construction. The Act states that whenever a home construction standard under the Act is in effect, no state "shall have any authority to establish . . . any standard regarding construction or safety applicable to the same aspect of performance of such manufacturered home which is not identical" to the federal standard.[26]

In *Shorter v. Champion Home Builders Co.*,[27] a homeowner's claim that particleboard used by the defendant homebuilder emitted hazardous levels of formaldehyde vapors, Champion claimed that its compliance with NMHCSSA standards preempted plaintiff's tort claims that the particleboard was defectively designed and that the manufacturer failed to provide adequate warnings of the risks of formaldehyde exposure. The federal trial court denied the home builder's motion for summary judgment, finding that the federal act and its attendant standards did not preclude a finding, under state law, that the manufacturer should have taken additional measures to lessen or eliminate risks. The court stated: "[t]he court cannot find, nor has Champion offered, any evidence in the legislative history of the Act that would suggest that a state law claim would frustrate the intent of Congress in reducing personal injuries in mobile homes. If anything, the availability of additional state law claims may serve to further reduce the number of personal injuries. . . . Thus, the Court finds that plaintiff's state law tort claim is not preempted by federal law."

14.4.2.3 *Cigarettes and Pharmaceuticals*

Active litigation surrounds the issue of whether the Cigarette Labeling and Advertising Act[28] impliedly preempts lawsuits, under state law, claiming that tobacco products are unreasonably dangerous and defective or that the manufacturers of these products have failed to provide adequate warnings as to the health hazards of smoking tobacco, chewing tobacco, or snuff products. The Labeling Act has been interpreted as impliedly preempting common law fraud claims against manufacturers of tobacco products,[29] and to date, most suits have resulted in summary disposition for the manufacturers, on the grounds that the Labeling Act evidenced congressional intent that compliance with labeling required by federal law preclude pursuit of claims that manufacturer warnings should have been more strongly worded or particularized.[30]

To be contrasted, many courts have suggested that pharmaceutical manufacturer satisfaction of Food, Drug and Cosmetic Act requirements for drug or vaccine licensure will not impliedly preempt tort claims under state law alleging improper manufacture,[31] with others stating that a pharmaceutical manufacturer's compliance with the requirements of the FDA will not preempt claims of failure to warn adequately or of defective formulation.[32]

14.5 ISSUE PRECLUSION

Issue preclusion, also known as collateral estoppel, prevents a party from relitigating an issue necessarily and finally decided in an earlier suit. To employ the defense of issue preclusion, the defendant must show that the issue in the current lawsuit, such as labeling adequacy or degree of hazard, is identical to the issue decided in an earlier trial.

Many obstacles are placed in the path of a proponent's proof of issue identity. A paradigm of the proponent's challenge can be found in the litigation following *Borel v. Fibreboard Paper Products Corp.*[33] In *Borel*, the plaintiff, an industrial worker who contracted asbestosis and mesothelioma, brought suit against a number of insulation manufacturers, alleging that the defendants were strictly liable for failing to warn about the dangers of long-term exposure to asbestos. Based on the evidence presented, the jury found the manufacturers were liable because they failed to warn about the foreseeable dangers associated with exposure to asbestos. Ten years later, different plaintiffs brought an action in which they sought to utilize *Borel* to preclude the litigation of the issue of causation. In *Hardy v. Johns-Manville Sales Corp.*,[34] the trial court held that *Borel* established as a matter of law that asbestos is a substance that can produce asbestosis and mesothelioma, and that no warnings were issued by any of the asbestos insulators prior to 1964. As a consequence, the trial court held, "the plaintiff need not prove that defendants either knew or should have known of the dangerous propensities of their products and therefore should have warned consumers of these dangers, defendants being precluded from showing otherwise."[35] The Fifth Circuit reversed, concluding that *Borel* did not necessarily decide the state of manufacturer knowledge as to the dangers of asbestos exposure.[36] Specifically, *Borel* did not resolve as a matter of fact that "all manufacturers of asbestos-containing insulation products had a duty to warn as of 1936, and all failed to warn adequately after 1964."[37] A proponent's failure to demonstrate issue commonality was also evidenced in *Walker v. GAF Corp.*,[38] where claimant's intentional tort action against an employer followed an adverse determination in plaintiff's negligence and strict products liability claims.[39]

NOTES

1. "Contributory negligence is conduct on the part of the plaintiff which falls below the standard to which he should conform for his own protection, and which is a legally contributing cause cooperating with the negligence of the defendant in bringing about the plaintiff's harm." Restatement, Second, Torts § 463.
2. 99 N.C. App. 275, 392 S.E.2d 761, Prod. Liab. Rep. (CCH) ¶ 12,565 (N.C. App. 1990).
3. _____ A.2d _____, Prod. Liab. Rep. (CCH) ¶ 12,562 (Md. Spec. App. 1989).
4. E.g., *Champs Convenience Stores, Inc. v. United Chemical Co., Inc.*, 99 N.C. App. 275, 392 S.E.2d 761, Prod. Liab. Rep. (CCH) ¶ 12,565 (N.C. App. 1990) (even where manufacturer erroneously delivered strong solvent rather than floor cleaning compound, employee's contributory negligence in not reading the name of the product or its instructions barred store's recovery under North Carolina Products Liability Act where application of cleaning solvent, rather than dusting product, resulted in damage to floor).
5. "A plaintiff who voluntarily assumes a risk of harm arising from the negligent or reckless conduct of the defendant cannot recover for such harm." Restatement, Second, Torts § 496A.
6. *Byrne v. S.C.M. Corp.*, 545 N.E.2d 105 (table) 127 Ill.2d 612, Prod. Liab. Rep. (CCH) ¶ 12,170 (Ill. App. 1989) (injurious exposure to fumes from epoxy paint).
7. E.g., *Richards v. C. Schmidt Co.*, 54 Ohio App. 3d 123, 561 N.E.2d 569 (Ohio App. 1989) ("The record contained no evidence that the warnings provided by Ashland were inadequate. . . . Instead, [plaintiff] stated that he read and understood the warnings, but that he ignored them. [Plaintiff] improperly used the chemicals to clean himself after work even though he read the warnings cautioning against repeated contact with skin. That improper use, contrary to clear warnings, removed the existence of any genuine issue of material fact relative to the alleged failure to warn." *Id.*, 54 Ohio App. 3d at 125, 561 N.E.2d at 572.).
8. 545 N.E.2d 105 (table), 127 Ill.2d 612 , Prod. Liab. Rep. (CCH) ¶ 12,170 (Ill. App. 1989).
9. 487 U.S. 500 (1988).
10. *Id.*, 487 U.S. at 512.
11. E.g., *Garner v. Santoro*, 865 F.2d 629 (5th Cir. 1989) (paint used in spraying destroyers built for the Navy); *In re Joint E. & S. Dist. New York Asbestos Litigation*, 897 F.2d 626 (2d Cir. 1990) (asbestos-based cement used in Navy yard).
12. *Jackson v. Deft, Inc.*, 223 Cal. App. 3d 1305, Prod. Liab. Rep. (CCH) ¶ 12,728 at 38,246 (1990) ("In our view, if a product is produced according to military specifications and used by the military because of particular qualities that serve a military purpose, and incidentally sold commercially as well, that product may still nonetheless qualify as military equipment under the military contractor defense.").

13. *Garner v. Santoro,* 865 F.2d 629 (5th Cir. 1989) (risks of inhalation of paint fumes); *In re Joint E. & S. Dist. New York Asbestos Litigation,* 897 F.2d 626 (2d Cir. 1990) (risks of exposure to asbestos-based cement).
14. *In re Joint E. & S. Dist. New York Asbestos Lit.,* 897 F.2d 626, 632 (2d Cir. 1990).
15. 223 Cal. App. 3d 1305, Prod. Liab. Rep. (CCH) ¶ 12,728 (1990).
16. *Id.,* ¶ 12,728 at 38,245. The court added: "Instead, the evidence suggested that no such limitation existed. For instance, Kopper's DiMaio testified that the government's specifications concerning labels were minimum standards, and that if the company considered it appropriate, it could add to that information without violating the specifications." *Id.* at 38,245.
17. 858 F.2d 712 (Fed. Cir. 1988).
18. 21 U.S.C. § 360k provides in part: "Except as provided in subsection (b) of this section, no state or political subdivision of a State may establish or continue in effect with respect to a device intended for human use any requirement — (1) which is different from, or in addition to, any requirement applicable under this chapter to the device, and (2) which relates to the safety or effectiveness of the device or to any other matter included in a requirement applicable to the device under this chapter."
19. See *Krause v. Kimberly Clark Corp.,* 749 F. Supp. 164, Prod. Liab. Rep. (CCH) ¶ 12,511 (W.D. Mich. 1990), where the court states: "After reviewing the applicable statutes and regulations, the court finds that state law claims are preempted only to the extent that they challenge the adequacy of warning or labeling on tampon packages." *Id.,* Prod. Liab. Rep. (CCH) ¶ 12,511 at 37,194–37,195 (collecting authority). The court found specifically that plaintiff's separate counts in negligence and breach of implied warranty were not preempted, and thus could, on the facts and pleadings, withstand a motion for summary judgment. *Id.,* Prod. Liab. Rep. (CCH) ¶ 12,511 at 37,195.
20. *Berger v. Personal Products, Inc.,* 115 Wash. 2d 267, 797 P.2d 1148 (Wash. 1990), *cert. denied* _____ U.S. _____ (1991) (state law action against tampon manufacturer alleging inadequate warnings and instructions).
21. U.S.C. § 136v(b).
22. _____ U.S. _____, 111 S.Ct 2476, 33 E.R.C.(BNA) 1265 (1991).
23. 763 F. Supp. 1500 (W.D. Mo. 1991).
24. *Professional Lawn Care Assoc. v. Milford,* 909 F.2d 929 (6th Cir. 1990).
25. 42 U.S.C. § 5401–5427.
26. 42 U.S.C. § 5403(d).
27. _____ F. Supp. _____, 1991 WL 212776 (N.D. Ohio 1991).
28. 15 U.S.C. §§ 1331–41.
29. E.g., *Rogers v. R.J. Reynolds Co.,* 557 N.E.2d 1045, Prod. Liab. Rep. (CCH) ¶ 12,572 (Ind. App. 1990).
30. See generally Note, "Inadequate Warning Claims Preempted by Cigarette Labeling Act: *Palmer v. Liggett Group. Inc.,*" 34 *Loyola Law Rev.* 419 (1988).
31. E.g., *Graham v. Wyeth-Ayerest Laboratories, Div. of American Home Products Corp.,* 906 F.2d 1399 (6th Cir.), *cert. denied,* _____ U.S._____ (1990).
32. *Wyeth-Ayerest Laboratories v. Graham,* 906 F.2d 1399 (6th Cir.), *cert. denied,* _____ U.S._____ (1990).
33. 493 F.2d 1076 (5th Cir. 1973), *cert. denied,* 419 U.S. 869 (1974).
34. 681 F.2d 334 (5th Cir. 1982).

35. *Id.*, 682 F.2d at 336–337.
36. *Id.*, 681 F.2d at 345.
37. *Id.* The court further explained that: "[O]ur opinion in *Borel* merely approved of the various ways the jury could have come to a conclusion concerning strict liability for failure to warn. We did not say that any of the specific alternatives that the jury had before it were necessary or essential to its verdict." *Id.*, 681 F.2d at 345.
38. No. 88-3380 (6th Cir. Sept. 25, 1989) (LEXIS, GenFed library, USAPP).
39. *Id.* ("We believe the district court improperly granted summary judgment on the basis of collateral estoppel because Walker's bald claim that Celotex, his former employer, committed an intentional tort does not necessarily raise the precise issue of whether or not Walker suffers from asbestosis, which was the issue raised and actually litigated in the prior proceeding." *Id.* at 4.). For an example highlighting the effect of the context of the earlier litigation upon the court's assessment of the opponent's earlier opportunity and incentive to litigate, see *McCarthy v. Johns-Manville Sales Corp.*, 502 F. Supp. 335 (D. Mass. 1980) (offensive issue preclusion was not applied against defendant manufacturers of asbestos products because they could not have foreseen the advent of mass asbestos litigation at the time the prior case was tried); see also *Clay v. Johns-Manville Sales Corp.*, 722 F.2d 1289 (6th Cir. 1983), *cert. denied*, 467 U.S. 1253 (1984). Where arguable distinctions exist between the issue previously litigated and the one for which a party seeks preclusion, one factor to be evaluated is whether there exists "a substantial overlap between the evidence or argument to be advanced in the second proceeding and that advanced in the first." Restatement (Second) of Judgments § 27 Comment c (1982).

CHAPTER 15

Limitations of Actions

CONTENTS

Limitations of Actions

15.1 GENERALLY

Statutes of limitation prescribe the time period within which a complaint must be filed after the underlying claim "accrues". In tort claims for personal physical injury or property damage, counsel's familiarity with the language and the application of statutes of limitations can make the difference between gaining trial of the claim to a factfinder or its pretrial dismissal for untimely filing.

With toxic tort claims particularly, there may be a substantial period of latency between the injurious exposure to a product or process. To avoid the unfairness of having an injured person's claim divest before the injury and its cause were fairly discernable, most jurisdictions have adopted a "discovery" analysis for determining when plaintiff's cause of action "accrues". The Supreme Court in *Urie v. Thompson*[1] stated the logic behind discovery rules, noting that in long latency cases barring a claim before a plaintiff could possibly know about an injury would operate as an involuntary waiver of plaintiff's right to just compensation for tortious injury.

Accordingly, under modern "discovery" rules, a claim does not accrue for limitations purposes until plaintiff knew, or in the exercise of reasonable care should have known, of the cause of her injury and its possible causal connection to defendant's product or conduct.[2] For choice of law purposes, there is general agreement that statutes of limitations are procedural in nature, not substantive, thereby freeing the forum court to apply the limitations period of the jurisdiction where the suit is filed.[3] Where, however, the place of accrual of a nonresident claimant's personal injury cause of action does not point clearly to one of two states, some courts have elected to borrow the shorter of the two pertinent statutes of limitations.[4]

This chapter discusses several current issues involving tort statutes of limitations, with particular reference to the interpretation and application of limitations periods with claim accrual keyed to plaintiff's actual or constructive "discovery" of their injury and its cause. After mention of the discrete issues posed by choice of tort or warranty remedies, wrongful death actions, revival statutes, and long-arm jurisdiction, there follows a review of decisions revealing the nature and quality of evidence required to show when the plaintiff alleging a long latency harm learned of, or with due diligence *should* have discovered, his injury and its probable cause.

15.1.1 Claims in Tort or Warranty

While a claim asserting personal physical injury or risk thereof sounds in tort, and perhaps also implied warranty, a claim asserting only economic harm, i.e., damage to a product or other property, but not posing an unreasonable risk of personal injury, may require application of a warranty limitations period.[5] In some jurisdictions discovery rules have been interpreted to apply only to cases of bodily harm, and not to claims involving damage exclusively to property. Prominent among such actions are claims brought by building owners against manufacturers and installers of asbestos insulation.[6] Significantly, in some jurisdictions municipalities have avoided dismissal for untimely actions by arguing a common-law immunity from limitations periods where the city sues in its municipal capacity to vindicate public rights. This avoidance of the limitations period has succeeded in suits brought by cities against producers of asbestos-containing products in public buildings.[7]

Where a tortious incident is fairly suited to both tort and warranty remedies, it can be filed under the law which has a longer limitations period. For instance, a paint purchaser's claim against a paint seller for breach of implied warranty of merchantability and fitness was governed by the four-year limitations period set out in the Uniform Commercial Code (U.C.C.),[8] even though the action could have been brought under the state's Product Liability Act, with its 2-year limitations period.[9]

15.1.2 Wrongful Death or Survival

Some courts have held that the limitations period is derivative of the decedent's claims had he or she lived to bring a personal injury action. For example, in *Russell v. Ingersoll-Rand Co.*,[10] the plaintiff's survivors brought an action against a company that supplied the decedent with equipment for his sandblasting career. In 1981 decedent discovered he had silicosis and commenced suit against several suppliers but not against the current defendant. Following his January 1988 death, decedent's survivors brought suit in March 1988 against the current defendant, only to have the court grant summary judgment for defendant, reasoning that had decedent not perished, his action against the current defendant would have been time barred as the injury was discovered in 1981. In other jurisdictions, however, a survivor's legal right to recover is considered independent from any claim that might have vested in the decedent prior to his death. Decisions following this rule hold that the statute of limitations for wrongful death actions begins to run upon the death of the decedent, not the time that the injury was discovered.[11]

15.1.3 Statutes of Repose

Whereas statutes of limitation set a time period in which a claim must be brought after it accrues, statutes of repose bar a claim altogether where suit is not brought within a time certain after the original sale of the toxic product. Unlike statutes of limitation, which "are generally seen as running from the time of injury, or discovery of the injury in cases where that is difficult to detect[,]"[12] statutes of repose have been invoked in asbestos litigation to bar plaintiff's claims of personal injury arising from exposure to defendant's products many years before. For example, in *Schultz v. Keene Corp.*,[13] plaintiffs husband and wife alleged that the husband's asbestosis was caused by exposure to defendant's asbestos products over 30 years prior to filing the claim. The trial court applied the Illinois 12-year statute of repose[14] to bar plaintiff's count in strict-products liability.[15]

In some jurisdictions, statutes of repose applicable to injuries arising from improvements to real property have been held inapplicable to the manufacturers of products used in the construction of those improvements. A Louisiana federal court so interpreted Oklahoma's statute of repose in a suit brought against the manufacturer of "Mono-Kote", a fireproofing substance containing asbestos.[16]

15.1.4 Revival Statutes

A revival statute is a procedural device that revives an otherwise time-barred remedy. Consequently a revival statute will not be interpreted to create a substantive cause of action not otherwise supported by a statute or the common law.

Constitutional challenges to revival statutes have fared variably.[17] Defendant product manufacturers have raised successful constitutional challenges to revival statutes where the statute has been interpreted as reviving actions previously barred by the applicable limitations period.[18] To be contrasted are other decisions upholding state revival statutes as a permissible exercise of the legislature's retroactive modification of a statute of limitations, and not an unlawful deprivation of a property right vested in a potential defendant.[19]

In *Anderson v. Eli Lilly & Co.*,[20] plaintiffs husband and wife asserted that the revival provision of New York's toxic tort statute[21] permitted their claim for loss of consortium. Specifically the plaintiffs argued that the wife's *in utero* exposure to DES prevented her from bearing natural children in wedlock. Finding a loss of consortium claim based on defendants' tortious conduct *prior* to the couple's marriage to be meritless, the court held that the procedural device provided by the revival statute did not, standing alone, create a loss of consortium claim.[22] Significantly, the enactment of a discovery rule will not revive an action already time barred. For example, in *Avery v. Shulman*,[23] evidence showed that a group of fire fighters was exposed to PCBs on July 24, 1982, and again on July 12, 1983, at a time when applicable law prescribed a 3-year limitations period. The fire fighters had until July 1985 to file their action, but did not do so until October of that year. In 1986 New York abandoned the date of injury rule in favor of the discovery rule. Having discovered their injuries in 1984, the fire fighters argued that the 1986 discovery statute should apply and the suit should be considered timely because they filed it only 1 year after discovery. The court disagreed, holding that the discovery rule is inapplicable to toxic tort cases where the tortious act occurred prior to July 1, 1986 and the injury is discovered before July 1, 1986.

15.1.5 Discovery Periods and Long-Arm Jurisdiction

In *Universal Engineering Corp. v. Perez*,[24] the Supreme Court of Florida recognized that occupational diseases are different from other personal injury actions. The court found that a cause of action accrues for statute of limitations purposes, not from the time of the infliction of the injury, but "from the time that the employee knows or should have known that the disease was occupational in origin, even though the diagnosis of the exact cause has not been made."[25] Consistently, legislatures in some states have amended their long-arm statutes to incorporate "discovery" based limitations periods for claims arising from long-latency occupational diseases.[26] For instance, in the consolidated action *In Re Asbestos Litigation*,[27] the federal trial court observed that the discovery statute of limitations contained in Florida's amended jurisdictional statute should apply to asbestos claims.

15.2 TOLLING STATUTES OF LIMITATIONS

15.2.1 Plaintiff's Actual and Constructive Awareness of Claim

Under a discovery interpretation of a limitations period, the plaintiff's claim accrues the date plaintiffs knew, or in the exercise of reasonable care should have known, of the nature of the harm and its causal connection to defendant's conduct.[28] Under Maryland law, for example, the limitations period begins to run when the plaintiff gains "knowledge of circumstances which ought to have put a person of ordinary prudence on inquiry."[29]

In some circumstances, a plaintiff's awareness of the substantial certainty that future harm will result is enough to trigger the running of the statute of limitations.[30] Statutes of limitations tied to plaintiff's "discovery" of the injury or loss are concededly more permissive than are so-called "date of injury" accrual rules. Nonetheless, a discovery rule should not permit a claimant to delay filing a claim until the extent of damages are "measurable with absolute certainty", for a limitations period may begin to run when a claimant is on notice of a compensable injury, though not necessarily of its extent.[31] Irrespective of the exact test applied under the discovery rule, the date of accrual of the plaintiff's cause of action is ordinarily a question of fact for the jury.[32]

While it is true that under a discovery interpretation limitations periods do not begin to run with the act that precipitates a latent injury, cases abound where circumstantial evidence established a plaintiff's prior notice of his injury, thereby time barring suits filed substantially thereafter. For example, in *Hess v. Firestone Plastics Co.*,[33] plaintiffs claimed that the defendants' landfill leaked chemicals which contaminated their drinking water. Interpreting its discovery rule in regard to its conventional 3-year limitations period for tort actions, the Maryland court noted several incidents that evidenced the plaintiffs' awareness of their claims more than 3 years prior to the filing of the present action. While plaintiffs' suit was filed March 25, 1988, the court credited evidence that by March 15, 1985, "[p]laintiffs were on notice that chemicals might be leaking out of the landfill into their water supply; [they] had filed two administrative complaints against one of the real estate agents involved in the purchase and sale of plaintiff's property; and, had retained an attorney to pursue possible legal action for personal injuries and economic damages."[34] Accordingly, the appeals court affirmed the trial judge's dismissal of the plaintiff's claim as time-barred.

Likewise, in one suit brought by a daughter whose mother had taken DES during pregnancy, the court ruled that the plaintiff's cause of action accrued 10 years before she filed suit and was therefore time-barred. In reaching this conclusion, the court relied on evidence that during an earlier period the claimant underwent a hysterectomy, met with and received the opinions of a medical researcher, and began collecting relevant magazine articles.[35]

Some jurisdictions have adopted the "reasonable possibility" test. Under this standard the statute of limitations begins to run when the plaintiff knew of or should have known of the "reasonable possibility" of the connection between the defendant's toxin and the claimant's injury or disease.[36] In determining when a plaintiff knew or had reason to know of the nature and causation of his injury particular reference will be made to the quality of the information available to him. Often times the circumstances in which the claimant received medical information, and specific medical observations, require the conclusion that the claimant cannot be fairly charged with actual or constructive knowledge of the nature of the injury at the time it happened. For example, in an asbestos-related injury suit, an appellate court concluded that a letter received by a claimant sent from his physician, referring to "some scarring" and "decrement" in his pulmonary function, was sufficiently innocuous so as not to put the claimant on notice of his occupational illness.[37] Even with the significance attached to the timing and nature of any medical opinion that a plaintiff may have received, it has been reiterated that a plaintiff is not entitled to an instruction that an asbestosis claim case accrues for limitations purposes upon the actual diagnosis of the illness.[38]

Cacciacarne v. G. D. Searle & Co.,[39] an intrauterine device (IUD) suit, provides another instructive example of what medical events will, or will not, be deemed to have sufficiently informed a plaintiff of the injury and its probable origin. While the appellate court held that the plaintiff's cause of action accrued when she was informed by her physician that she could not conceive, the court also emphasized that the earlier statement of another physician that her IUD might be the cause of her tubal blockage was insufficient to put her on notice of her claim, particularly where the evidence showed that within a month of the earlier conversation both the plaintiff and her first physician concluded that her tubes were no longer blocked and that she could conceive.[40]

Another illustrative case about the level of medical knowledge sufficient to trigger the running of the statute of limitations is *University of Miami v. Bogorff.*[41] In that products liability case against the manufacturer of the drug methotrexate, the plaintiff's 3-year-old son received an intrathecal (cerebrospinal) injection for his leukemia. Sometime thereafter the boy suffered quadriplegia and severe brain damage. Reversing the appellate court's opinion that the limitations period had not expired because the claim did not accrue until long after the child's last injection, the Florida Supreme Court held that statute of limitations began to run when the parents became aware of the boy's injuries and of the possible involvement of methotrexate. The Supreme Court reasoned that although the injuries were not readily distinguishable from the effects of leukemia on the boy, his parents were aware that their child's condition changed dramatically after the last injection and had constructive knowledge of medical opinion that the methotrexate may have contributed to the injuries.

In the context of asbestos-related disease, most courts recognize that a plaintiff having the illness of asbestosis does not presage illness of another disease such as mesothelioma. Consistent with this, courts in states employing the discovery rule have held that a wrongful death action seeking damages for a worker's death by mesothelioma accrued when the worker was or reasonably could have been diagnosed as having mesothelioma, and not years before upon the worker being diagnosed as having a different asbestos-related disease.[42] Likewise, a worker's diagnosis that he suffered from contact dermatitis as a result of exposure to the defendant's chemicals did not operate as an accrual of his subsequent and later diagnosed claim for mycosis fungoides, a form of skin cancer.[43] In contrast to these asbestos-related cases is a zinc-related case. The discovery by workers that their multiple sclerosis, which had previously manifested itself, was again caused to manifest itself or become accelerated or aggravated from exposure to zinc products sold by the defendant's employer to the workers' employer was held not a separate and distinct disease so as to trigger a new limitations period.[44]

There are situations in which a sufficiently "sudden and traumatic" injury of nonapparent origin is enough to put the injured party or their agent on notice. In *Kempfer v. Shiley Inc.,*[45] the court held that decedent's survivors' wrongful death claim arose upon decedent's death, which occurred 4 months after his cardiac arrest 2 months after receiving a valve implantation. Although the evidence in that suit showed that the survivors did not learn that the valve may have been defective until several years after the death, the court held that decedent's cardiac arrest in relative proximity to the implantation put the survivors on notice of the potential causal connection between the arrest and the design or manufacture qualities of the valve.[46]

A plaintiff's earlier receipt of cautionary information from the manufacturer or seller of a product may be probative of the level of plaintiff's knowledge regarding the nature of a later injury and its causal connection to defendant's conduct. A manufacturer's notice to plaintiff as to potential product risks or performance limitations may be held sufficient to have informed plaintiff of the existence and nature of his potential claim.[47]

When a corporate subsidiary receives notice of a claim before the parent company does, the notice may be deemed to inform the company of the claim as if it had received the information of the claim at the same time its subsidiary did. Under the so-called "relation back" rule, a manufacturer of methyl isocyanate was held to have received notice of a claim for alleged injuries caused by that substance upon the receipt, by its registered agent for service of process, of the complaint, even where the complaint named a subsidiary of the manufacturer that was created to hold and license the trademark.[48] In other circumstances, an amendment of a complaint to add new parties defendant and new causes of action will be barred by the applicable limitations period.[49] A court's adoption of a theory of market share or related alternate liability[50] should not, however, be interpreted as uniting or allying defendant's interests as to trigger the relation back doctrine to preserve a claim that is otherwise untimely.[51]

NOTES

1. 337 U.S. 163, 169 (1949).
2. E.g., "[N]o claim under this [products liability] chapter may be brought more than three years after the time the claimant discovered, or in the exercise of due diligence should have discovered, the harm and its cause." Rev. Code Wash. 7.72.060(3).
3. *Menne v. Celotex Corp.*, 722 F. Supp. 662, Prod. Liab. Rep. (CCH) ¶ 12,468 at 36,990 (D. Kan. 1989).
4. E.g., *Allen v. Hanszer*, 560 N.Y.S.2d 593 (N.Y. Sup. 1990).
5. E.g., *Drew v. United Producers and Consumers Cooperative*, 778 P.2d 1227 (Ariz. 1988) (en banc) (suit against pig feed seller).
6. E.g., *Corporation of Mercer University v. National Gypsum Co.*, 877 F.2d 35 (11th Cir. 1989).
7. E.g., *District of Columbia v. Owens-Corning Fiberglas Corp.*, 572 A.2d 394 (D.C. App. 1989).
8. See U.C.C. § 2-725.
9. *B & B Paint Corp. v. Shrock Mfg., Inc.*, 568 N.E.2d 1017 (Ind. App. 1991).

10. *Russell v. Ingersoll-Rand Co.*, 795 S.W.2d 243 (Tex. App. 1990). See also *Dunn v. Pacific Employers Ins. Co.*, 400 S.E.2d 63 (N.C. App. 1991) (action time-barred even though brought within 2 years of a worker's death allegedly caused by exposure to hazardous chemicals at work).
11. *Frongillo v. Grimmett*, 788 P.2d 102 (Ariz. App. 1989) (held: wrongful death action accrued on the date of the husband's death even though the statute of limitations on the husband's personal injury action had expired before his death); see also *Moreno v. Sterling Drug Inc.*, 787 S.W.2d 348 (Tex. 1990) (Texas law). Compare *Conner v. George W. Whitesides Co.*, 1990 WL 134962 (Ky. App. 1990) (Kentucky discovery rule does not apply in wrongful death actions, including actions where death resulted from exposure to toxic chemicals).
12. *Menne v. Celotex Corp.*, 722 F. Supp. 662, Prod. Liab. Rep. (CCH) ¶ 12,468 at 36,990 (D. Kan. 1989) (citation omitted).
13. 729 F. Supp. 609, Prod. Liab. Rep. (CCH) ¶ 12,518 (N.D. Ill. 1990).
14. Ill. Rev. Stat. Ch. 110, ¶ 13-213(b) (1987), which requires that products liability claims based upon the doctrine of strict liability be commenced "within 12 years from the date of the first sale, lease or delivery of possession by a seller or 10 years from the date of the first sale, lease or delivery of possession to its initial user, consumer, or other nonseller, whichever period expires earlier."
15. *Schultz v. Keene Corp.*, 729 F. Supp. 609, Prod. Liab. Rep. (CCH) ¶ 12,518 (N.D. Ill. 1990). The court explained: "Much like the plaintiff's husband in *Blazek* [v. Nicolet, Inc., 173 Ill. App. 3d 324, 527 N.E.2d 568, 572 (Ill App. 1988)], John Schultz was first exposed to asbestos in the 1950s — approximately thirty years prior to the commencement of the instant case. Having filed suit on July 25, 1986, plaintiffs' claim clearly falls outside of the 12-year period of repose." *Id.*, 729 F. Supp. at 613.
16. *Uricam Corp., N.V. v. W. R. Grace & Co.*, 739 F. Supp. 1493 (E.D. La. 1990).
17. In some jurisdictions, defendant product manufacturers have successfully challenged the constitutional validity of revival statutes that have been interpreted as reviving actions previously barred by the applicable limitations period. E.g., *Waller v. Pittsburgh Corning Corp.*, 742 F. Supp. 581 (D. Kan. 1990) (worker claiming injurious exposure to asbestos products could not, via revival statute, maintain an action more than 10 years after last exposure; 10-year statute of repose expired prior to enactment of revival statute legislation).

    Other decisions have upheld revival statutes as permissible legislative modification of a statute of limitations, and not an unlawful deprivation of a property right vested in a potential defendant. E.g., *Independent School Dist. No. 157 v. W.R. Grace & Co.*, 752 F. Supp. 286 (D. Minn. 1990) (asbestos contamination).
18. E.g., *Waller v. Pittsburgh Corning Corp.*, 742 F. Supp. 581 (D. Kan. 1990) (worker claiming injurious exposure to asbestos products could not, via revival statute, maintain an action more than 10 years after last exposure; 10-year statute of repose expired prior to enactment of revival statute legislation).
19. *Independent School Dist. No. 157 v. W. R. Grace & Co.*, 752 F. Supp. 286 (D. Minn. 1990) (asbestos).
20. 158 A.D.2d 91, 557 N.Y.S.2d 981 (1990) motion for appeal granted 568 N.Y.S.2d 347.
21. L 1986, ch 682, Sect. 4.

22. *Anderson v. Eli Lilly & Co.*, 158 A.D.2d 91, 557 N.Y.S.2d 981 (1990). In another suit challenging the same New York revival statute, a state appellate court held that the 1-year time period for initiation actions was a statute of limitations, and not a condition precedent, and could, therefore, be extended by the time periods provided for in [separate] statutes treating delivery of summons and complaint and a 60-day period within which to effect personal service on defendants. *Clark v. Abbott Laboratories,* _____ N.Y.S.2d _____, 1990 WL 292022 (App. Div. 1990).
23. 566 N.Y.S.2d 785 (App. Div. 1991).
24. *Id.*
25. *Id.*
26. E.g., *In re Asbestos Litigation,* 679 F. Supp. 1094, (S.D. Fla. 1988) (Florida law). In that consolidated action, the trial court observed that the discovery statute of limitations contained in that state's amended jurisdiction statute should apply to asbestos claims, which are, in the court's words, "unique in products liability actions." The court stated that claims involving occupational diseases "are different from other personal injury actions." In determining when a cause of action accrues for statute of limitations purposes, the court found that the statute runs not from the time of infliction of the injury, but "from the time that the employee knows or should have known that the disease was occupational in origin, even though diagnosis of the exact cause has not been made." (citation omitted).
27. *Id.*
28. See note 2, supra.
29. *Hess v. Firestone Plastics Co.*, 872 F.2d 417, 29 E.R.C. (BNA) 1525, 1526 (4th Cir. 1989) (citation omitted). Applying Maryland law to a common law groundwater contamination suit, the court in *Hess* continued: "As of that date, the claimant 'is charged with knowledge of facts that would have been disclosed by a reasonably diligent investigation.'" *Id.* (citation omitted).
30. See *Brower v. E.I. DuPont de Nemours and Co.*, 792 P.2d 345 (Idaho 1990) (farmer's knowledge, through notification by herbicide manufacturer, of chemical dissipation/contamination problem deemed sufficient for plaintiff's awareness of his claim even though the product deficiency did not interfere with planting until a later time).
31. *Brower v. E. I. Dupont de Nemours & Co., id.* In that suit, brought by a farmer against the manufacturer of a herbicide claimed to have contaminated land, rendering it unsuited to planting a variety of crops, the court concluded that plaintiff's claim was time barred. The court stated: "[Plaintiffs'] damages may not have been measurable with absolute certainty in the fall of 1982, but their extent was capable of proof namely the proceeds to leasing land for potato crops over a four year period against the yield from wheat or barley." 792 P.2d at 347.
32. *Stark v. Celotex Corp.*, 58 Wash. App. 940, 795 P.2d 1165 (Wash. App. 1990) ("The point at which a plaintiff discovers or should have discovered that he or she is harmed and its cause is a question of fact for the jury to decide." *Id.* at 1166).
33. 872 F.2d 417, 29 E.R.C. (BNA) 1525 (4th Cir. 1989).

34. *Id.*, 29 E.R.C. (BNA) at 1526–1527 ("As the district court recognized, more than three years before filing this action, plaintiffs' attorney was prepared to present claims to Cecil County officials raising issues 'substantially identical' to those asserted in the instant case.").
35. *Bowen v. Eli Lilly & Co.*, 408 Mass. 204, 557 N.E.2d 739 (Mass. 1990).
36. E.g., *Evenson v. Osmose Wood Preserving Co. of America, Inc.*, 899 F.2d 701 (7th Cir. 1990).
37. *Wanner v. Philip Carey Mfg. Co.*, 580 A.2d 734 (N.J. App. 1989). In *Wanner* the court also took into account the fact that the letter from the physician arrived some considerable time after the examination, a circumstance that might further suggest to the patient the benign nature of the doctor's conclusions.
38. *Stark v. Celotex Corp.*, 58 Wash. App. 940, 795 P.2d 1165 (Wash. App. 1990).
39. 908 F.2d 95 (6th Cir. 1990).
40. *Id.*, 908 F.2d 95 (6th Cir. 1990) (Held: Ohio 2-year limitations period did not bar action). See also *Evenson v. Osmose Wood Preserving Co. of America*, 899 F.2d 701 (7th Cir. 1990) (held: wood treatment worker neither knew nor had reason to know of reasonable possibility that work-related exposure to chromated copper arsenate was cause of his injuries at any time prior to 2 years preceding his filing of complaint).
41. 583 So. 2d 1000, 16 Fla. L. Week. 149 (Fla. 1991).
42. *Potts v. Celotex Corp.*, ____ Tenn. ____, 796 S.W.2d 678 (Tenn. 1990) (Tennessee law).
43. *Zumo v. R.T. Vanderbilt Co., Inc.*, 527 So. 2d 1074 (La. App. 1988) (second diagnosis more than 3 years after diagnosis for dermatitis).
44. *Michael v. Ametelco Inc.*, 568 N.Y.S.2d 1003 (S.D.N.Y. 1991).
45. *Kumpfer v. Shiley, Inc.*, 741 F. Supp. 738 (N.D. Ill. 1990) (cardiac arrest 4 months after implantation of heart valve).
46. *Id.*
47. E.g., *Brower v. E. I. DuPont de Nemours & Co.*, 117 Idaho 780, 792 P.2d 345 (Idaho 1990) (herbicide manufacturer's notification to farmer that chemical's dissipation rate would be longer than anticipated, although the product problem did not prevent planting until a later date.)
48. *Peterson v. Sealed Air Corp.*, 902 F.2d 1234 (7th Cir. 1990).
49. E.g., *Christopher v. Duffy*, 556 N.E.2d 121 (Mass. App. 1990) (mother whose daughter died following hospitalization for lead poisoning could not, 6 years after filing initial suit against owners of apartment and contractor engaged in lead paint removal, amend complaint to state new causes of action against lead manufacturers).
50. Market share liability, alternate or alternative liability, and related theories are previously discussed in Chapter 2.
51. *Korren v. Eli Lilly & Co.*, 568 N.Y.S.2d 670 (N.Y. Sup. 1990) (adoption of market-share liability against DES manufacturers).

CHAPTER 16

Jurisdiction and Conflicts of Law

CONTENTS

Jurisdiction and Conflicts of Law

16.1 GENERALLY

In any toxic tort claim, the court must have a proper jurisdictional basis for exercising judicial power over the parties. Jurisdictional disputes most commonly arise over whether the court of one state can require a defendant to defend a toxic tort claim brought in that state. For example, if a truck owned by a New Mexico hauler of toxic wastes were to overturn on an Arizona road, contaminating the farmland of an Arizona resident, the Arizona plaintiff might bring suit in nuisance or trespass against the owners of the hauler. If the suit were filed in an Arizona court, it would create the question of whether the defendant, a New Mexico resident, should be required to appear before an Arizona court to defend the claim. Similar jurisdictional questions might arise in a suit brought against a California manufacturer of residential paneling who is sued in Michigan by a Michigan purchaser claiming injurious exposure to formaldehyde vapors.

While the complainant filing a complaint in a particular state or federal court presumably accepts the court's personal, or *in personam*, jurisdiction over the claim and the parties, a single defendant or multiple defendants may claim that the suit, if it is to be brought at all, must be tried in a different court. The issue is described as an evaluation of which states defendant can be deemed "present in" for purposes of service of legal process.

Every state, by statute or rule, defines the personal jurisdiction of its courts over residents, and persons or businesses that either do business in the forum state, or whose activities external to that state affect persons within the forum state. Businesses that are not incorporated in a particular state are often described as "foreign" corporations.

16.1.1 Personal Jurisdiction and Due Process

Bringing a person or a business before a court to answer a toxic tort claim is seldom challenged where defendant is served process in the state of its coporate domicile or in the state of its principal place of business. Nonresident defendants, however, routinely challenge suits brought against them with the argument that the court has no personal jurisdiction over them in that state and on that particular claim. A court that asserts *in personam* jurisdiction improperly over a nonresident defendant denies due process of law and renders any proceedings or judgment invalid.

In *International Shoe Corp. v. Washington,*[1] the Supreme Court established the rule that a state or federal court in any jurisdiction may assert jurisdiction over a particular "foreign" corporation only where that nonresident corporation has "certain minimum contacts with [the forum] such that the maintenance of the suit does not offend 'traditional notions of fair play and substantial justice'."[2] The later decision in *Hanson v. Deckla*[3] elaborated that the proper exercise of personal jurisdiction over a foreign corporation requires "that there be some act by which the defendant purposefully avails itself of the privilege of conducting activities within the forum state, thus invoking the benefits and protections of its laws."[4]

16.2 LONG-ARM JURISDICTION

Many states have enacted so-called "long-arm" statutes that define circumstances in which jurisdiction may be validly asserted over a nonresident defendant. The Illinois statute, representative of many, provides that jurisdiction may be exercised as to any suit arising from defendant's (a) "transaction of any business within this State;" (b) "commission of a tortious act within this State;" (c) "ownership, use or possession of any real estate situated in this State;" or (d) "[c]ontracting to insure any person, property or risk located within this State at the time of contracting."

In toxic tort claims as in other civil litigation, a two-prong analysis is employed in determining whether a court has personal jurisdiction over a party. The court first determines if the party and the claimed conduct of that party fit within the provisions of the forum state's long-arm jurisdiction statute.[5] Secondly, if the party fits within the long-arm statute, "the court must then consider whether the exercise of personal jurisdiction over the nonresident defendant offends notions of due process and fairness."[6]

Generally, where a defendant's contacts with the forum state are unrelated to the injury claimed, a court nevertheless may exercise "general personal jurisdiction" where the defendant has had "continuous and systematic" business contacts within the state. Specific personal jurisdiction, on the other hand, is based upon "a party's contacts with the forum state that are related to the litigation."[7] In the example mentioned above, of a New Mexico toxic-waste hauler and an Arizona plaintiff, a court might hold that the Arizona court had jurisdiction over the nuisance and trespass claims, because those harms occurred in Arizona, but not over other civil claims, not related to the contaminating spill, that plaintiff might wish to lodge against defendant.

A characteristic provision of state long-arm jurisdiction provides that the forum may exercise *in personam* jurisdiction over a foreign defendant whose "act or omission outside the state" causes a tortious injury within the state "if the tort-feasor regularly does or solicits business, or engages in any other persistent course of conduct, or derives substantial revenues from goods used or consumed or services rendered in this state[.]"[8] It has been held that "merely plac[ing] its product into the stream of commerce" should not suffice to satisfy the "minimum contacts" required for exercise of personal jurisdiction over a foreign defendant.[9] One federal appellate court has adopted an exacting five-factor analysis for minimum contacts[10] and expresses reluctance to confer *in personam* jurisdiction over a foreign seller that has neither shipped nor sold products directly to the forum state.[11]

There is conflicting authority concerning the effect of a business predecessor's activities upon permissible *in personam* jurisdiction over the successor. In one suit, the court found that *in personam* jurisdiction would properly lie for a successor in any jurisdiction that could have asserted personal jurisdiction over the predecessor.[12] Conversely, in another suit brought in another jurisdiction, the court held that a corporate successor's acquisition of its predecessor's product lines and its subsequent licensing of other companies to manufacture and distribute products generally within the U.S. will not support *in personam* jurisdiction in a jurisdiction where the successor itself neither conducted business nor targeted marketing.[13]

The long-arm statutes of some jurisdictions have been interpreted to authorize *in personam* jurisdiction in the forum court on plaintiff's showing that defendant manufacturers conspired to conceal the hazards of a toxic product.[14]

16.3 PROCEDURAL CHALLENGES

16.3.1 Class Actions

Certification of a claimant class action has been held warranted in diverse asbestos-related products liability suits where the court finds that the finite financial assets of one asbestos manufacturer created a substantial risk of compromising payment of past and prospective liability judgments.[15] Courts in some jurisdictions have found that "varying damages" and "individualized . . . compensation" should not preclude class certification where a large number of plaintiffs "seek to recover for injuries which resulted from a common course of conduct."[16]

In one suit, brought by residents of Port Richmond, PA, claiming that a lead processor released hazardous amounts of lead into the regional environment, the trial judge found that plaintiffs satisfied the five requisites for class-action certification in that jurisdiction: numerosity (123 plaintiffs), commonality, typicality, adequacy of representation, and the fairness and efficiency of the class-action resolution of the action.[17]

In the context of Agent Orange litigation, some courts have suggested that class-action treatment of the indeterminate plaintiff achieves measurable "judicial and economic efficiencies."[18] An argument against facilitating class-action litigation of the mass toxic tort case is that "making it easier to sue may increase the possibility of obtaining unjustified settlements by a form of legal coercion."[19]

16.3.2 Forum *Non Conveniens*

Even where personal jurisdiction may be validly exercised over a nonresident defendant, the court is not compelled to proceed. Defendant may claim that the forum is so inconvenient, and poses such burdens upon it, the witnesses, and the judicial process, that in fairness the trial should be conducted in another jurisdiction. The doctrine involved is called forum *non conveniens* (inconvenient forum), and to prevail, the moving party must by affidavit or other factual proof demonstrate that public and private hardships commend trial in another forum.[20]

In *Gulf Oil Co. v. Gilbert,*[21] the Supreme Court listed the following criteria as pertinent to the trial court's discretion as to the fairness of trying the suit where filed, or elsewhere: "relative ease of access to sources of proof; availability of compulsory process for attendence of unwilling, and the cost of obtaining attendance of willing, witnesses; possibility of view of premises, if view would be appropriate to the action; and all other practical problems that make trial of a case easy, expeditious, and inexpensive."[22]

16.4 CONFLICTS OF LAW

A court will not always apply the rules of evidence, liability, or damages of the state where the suit is brought. State and federal trial courts alike may conclude that even though it was permissible for plaintiff to sue defendant in one state, the injury, the transactions giving rise to the injury, or other interests of the parties bear a closer relationship to another jurisdiction.[23]

Moreover, a federal court is not bound to apply a state procedure for dismissal of certain codefendants where the court concludes that the state statute is procedural.[24] Conventional statutes of limitations are considered procedural in nature, serving as "instruments of public policy and of court management," rather than "confer[ring] upon defendants any right to be free from liability[.]"[25] Accordingly, in many jurisdictions, *lex fori* (the law of the forum), controls determination of the applicable limitations period.[26] Generally, in these situations the court will follow the procedural law of the jurisdiction in which it sits, including applicable statutes of limitations, and the substantive law of the state where the cause of action arose. Statutes of repose, on the other hand, are considered substantive rules of law, intended as "a substantive definition of rights as distinguished from a procedural limitation on the remedy used to enforce rights."[27]

16.4.1 *Lex Loci Delicti* and Modern Interest Analysis

In many jurisdictions, courts will be presumed to apply the substantive law of the place where the injury occurred, a doctrine known as *lex loci delicti.* Even in a state adhering to the rule of *lex loci delicti,* a court may elect to apply the law of its own state where a state concludes that *its* law demonstrates such a strong public policy commitment to a claim or a defense that the state's interest in application of that law in its courts transcends the competing interests of the state in which the injury occurred. A Tennessee federal court so concluded in *Trahan v. E.R. Squibb & Sons,*[28] in choosing to apply Tennessee's law of strict products liability to the claim of a woman alleging that she developed cervical cancer resulting from her mother's ingestion of DES.

Courts employing the "most significant relationship" approach to choice of law most often refer to the criteria stated in Restatement, Second, Conflict of Laws §§ 6 and 145. The restatement invites courts to evaluate the following types of "contacts" the claim may have with the respective jurisdictions: "(a) the place where the injury occurred; (b) the place where the conduct causing the injury occurred; (c) the domicile, residence, nationality, place of incorporation and place of business of the parties; and (d) the place where the relationship, if any, between the parties is centered."

No one criterion dominates the court's consideration. Rather, in deciding which law to apply, the court will evaluate the entire context of the claim, and the relationship of a party to other parties and to the one or more states with which the claim has some nexus. In the context of a modern and mobile workforce and population, the criterion of "where the injury occurred" can often not be given much weight, as it is often impossible to ascertain. In such circumstances, a court will give heightened consideration to the question of the place where defendant's conduct occurred.[29]

An instructive application of the most significant relationship test is found in *Elmore v. Owens-Illinois, Inc.*,[30] a suit brought by an asbestos worker who was exposed for years to asbestos products designed to insulate pipes, boilers, turbines, and like products. The substantive law of Kansas was quite favorable to defendant, as it would premise liability upon plaintiff's showing that "at the time defendant sold Kaylo [it] knew, or in the exercise of reasonably developed human skill and foresight could have known, that Kaylo was unreasonably dangerous to insulation workers and that an alternative safe design was technically feasible[.]"

Applying a "most significant relationship" test, the appeals court upheld the trial judge's decision that Missouri law should apply. In so doing, the court noted that during the 10-year period in which defendant manufactured the product, Mr. Elmore worked principally for Missouri-based employers; that the amount of Social Security paid by Missouri employers was 17 times as great as that contributed by Kansas employers; and that he was for this time a member of a Missouri asbestos worker union. On the basis of this evaluation, the appeals court concluded, "'the place where the relationship, if any, between the parties is centered' appears to have the greatest relative importance with respect to the particular issue. That place was Kansas City, Missouri. Plaintiff Elmore's injury was intimately and inextricably involved with his employment[, and] his employment was chiefly for Kansas City-based employers[.]''

NOTES

1. 326 U.S. 310 (1945).
2. *Id.*, 326 U.S. at 316.
3. 357 U.S. 235 (1958).
4. *Id.*, 357 U.S. at 250-251.
5. E.g., *Bond v. Octagon Process, Inc.*, 745 F. Supp. 710, Prod. Liab. Rep. (CCH) ¶ 12,737 at 38,281 (M.D. Ga. 1990), a suit brought by plaintiff alleging that he contracted Raynaud's phenomenon from contact with defendant's cleaning solvent. The trial court, before dismissing for absence of *in personam* jurisdiction, observed: ''Plaintiffs must show that the defendant is amenable to a suit under this statute [O.C.G.A. Sect. 9-10-91 (3)] before any federal constitutional questions are reached.''
6. *Adolf v. A.P.I., Inc.*, 726 F. Supp. 764, Prod. Liab. Rep. (CCH) ¶ 12,497 at 37,127 (D.N.D. 1989) (citations omitted).
7. *Bond v. Octagon Process, Inc.*, 745 F. Supp. 710, Prod. Liab. Rep. (CCH) ¶ 12,737 at 38,281–38,282 n.1 (M.D. Ga. 1990) (Reynaud's phenomenon allegedly resulting from plaintiff's contact with defendant's cleaning solvent).
8. *Bond v. Octagon Process, Inc., id.*, Prod. Liab. Rep. (CCH) ¶ 12,737 (M.D. Ga. 1990) (citing O.C.G.A. § 9-10-91).
9. *Bond v. Octagon Process, Inc.*, 745 F. Supp. 710, Prod. Liab. Rep. (CCH) ¶ 12, 737 at 38,282–38,283 (M.D. Ga. 1990) (''Even assuming that the plaintiffs are correct when they claim that Octagon knew that its product would eventually enter Georgia . . . *Asahi Metal Ind. v. Superior Court*, 480 U.S. 102 (1987) made it clear that this simply is not enough to form sufficient minimum contacts so that Octagon could foresee being haled into court here. . . . When Octagon sold the solvent to the Department of Defense, which acted as an independent distributor, it merely placed its product into the stream of commerce, an act which alone will not allow this court to exercise personal jurisdiction in Georgia.'').
10. *Adolf v. A.P.I., Inc.*, 726 F. Supp. 764, Prod. Liab. Rep. (CCH) ¶ 12,497 at 37,128 (D.N.D. 1989) ''(1) The nature and quality of the contacts with the forum state; (2) the quantity of contacts with the forum state; (3) the relation of the cause of action to the contacts; (4) the interest of the forum state in providing a forum for its residents; and (5) the convenience of the parties.''
11. *Id.*, Prod. Liab. Rep. (CCH ¶ 12,497 at 37,129) ''Since this court is bound by the Eighth Circuit's views . . . the fact that defendant never shipped or sold any [of its asbestos] products directly to North Dakota is controlling. The Eighth Circuit has rejected the stream of commerce theory of jurisdictions, and LAQ cannot be found to have consented to jurisdiction by its acts.'' *Id.* (citation omitted).

12. *Simmers v. American Cyanamid Corp.*, 576 A.2d 376 (Pa. Super. 1990) (suit involving radioactive contrast dye Thorotrast).
13. *Ward v. Armstrong World Indus. Inc.*, 677 F. Supp. 1092, Prod. Liab. Rep. (CCH) ¶ 11,757 (D. Colo. 1988).
14. E.g., *In re North Dakota Personal Injury Litigation No. 1*, 737 F. Supp. 1087 (D.N.D. 1990) (jurisdiction over Canadian supplier of raw asbestos).
15. *In re Joint Eastern and Southern Districts Asbestos Litigation*, 132 F.R.D. 332 (E.D.N.Y. 1990).
16. *Wagner v. Anzon, Inc.*, ____ A.2d ____, 1990 Toxics Law Rptr. (BNA) 398 (August 22, 1990) (Pa. Ct. Comm. Pls., July 31, 1990).
17. The court noted specifically that plaintiffs' counsel and the proposed class representatives "have conducted themselves in a manner which suggests that there is no reason to fear that these parties have any conflicting interest in the instant case or that any such problem will later arise." *Id.*
18. *In re 'Agent Orange' Product Liability Litigation*, 597 F. Supp. 740 (E.D.N.Y. 1984) ("Although recent studies suggest that defendants in mass tort cases spend more money than plaintiffs do for legal services . . . , [D]efendants, often large corporations, are usually better able to bear the cost of litigating individual suits than are plaintiffs' attorneys who are generally financing the litigation out of their own pockets. A class action, by allowing plaintiffs' attorneys to pool their resources and by forcing defendants to litigate all of plaintiffs' claims at once, should not only reduce the overall cost of the litigation significantly but should tend to equalize the odds between the two sides. A class action would also facilitate settlement by allowing defendants to buy their peace with all of the plaintiffs at once."). *Id.*, 597 F. Supp. at 841–842.
19. *In re "Agent Orange" Product Liability Litigation*, 597 F. Supp. 740 (E.D.N.Y. 1984).
20. E.g., *Picketts v. International Playtex, Inc.*, 215 Conn. 490, 576 A.2d 518 (Conn. 1990) (Canadian tampon manufacturer and American parent corporation failed to show that toxic shock action should be tried in British Columbia, even where action was brought by a Canadian woman in Canada, and plaintiff was treated in Canada; held: manufacturer did not support claim of hardship occasioned by lack of compulsory process in Connecticut for Canadian witnesses).
21. 330 U.S. 501 (1947).
22. *Id.*, 330 U.S. at 508.
23. Thus, for example, in *Smith v. Walter C. Best, Inc.*, 756 F. Supp. 878 (W.D. Pa. 1990), the federal trial court in Pennsylvania, hearing a foundry worker's suit against suppliers of silica dust, elected to apply not Pennsylvania law, but rather Ohio law.
24. *Pruett v. Goldline Laboratories, Inc.*, 751 F. Supp. 1372 (W.D. Mo. 1990) (in L-Trytophan suit, court interpreting state statute allowing "seller in the stream of commerce" to achieve dismissal from products liability action where total recovery could be had from another defendant).
25. *Goad v. Celotex Corp.*, 831 F.2d 508, 511 (4th Cir. 1987), cert. denied, ___ U.S.____, 108 S. Ct. 2871 (1988) (citations omitted).
26. E.g., *Menne v. Celotex Corp.*, 722 F. Supp. 662, Prod. Liab. Rep. (CCH) ¶ 12,468 at 36,988–36,989 (D. Kan. 1989) (Kansas law).
27. *Menne v. Celotex Corp.*, supra, Prod. Liab. Rep. (CCH) ¶ 12,468 at 36,990 (D. Kan. 1989) (citations omitted).

28. 567 F. Supp. 505, Prod. Liab. Rep. (CCH) ¶ 9827 (D. Tenn. 1983).
29. Restatement, Second, Conflict of Laws § 145, comment e.
30. ——— Mo. App. ———, Prod. Liab. Rep. (CCH) ¶ 9729 (Mo. App. 1983).

CHAPTER 17

Indemnity and Contribution

Indemnity and Contribution

17.1 INDEMNITY — GENERALLY

In toxic tort litigation a defendant found liable for damages may, under certain circumstances, seek indemnification from another for the amount of any judgment entered in a tort claim brought against it or for the amount paid out in settlement of the claim. Indemnity obligations may be explicitly stated, or they may be implied on the basis of the respective roles played by the parties in the distribution or use of the product or substance.

State law typically provides that intratortfeasor indemnification is available where (1) a contract between the parties explicitly so provides or (2) where a special legal relationship between the parties imports an implied right to indemnification.[1] Under principles of either express or implied indemnity, the indemnitee's active intervening negligence may bar its indemnification claim.[2]

Some express agreements of indemnification are general in their terms. One such agreement, construed in *Price v. Shell Oil Co.*,[3] provided that "[l]essee shall indemnify Shell against any and all claims and liability for injury or death of persons or damage to property caused by or happening in connection with the equipment or the condition, maintenance, possession, operation or use thereof."[4] The court in that suit held that this language was inadequate to require the lessee to indemnify Shell for either its negligence or for its strict products liability in placing a defective article on the market.

Most indemnifications claims are based on the argument that one party has an *implied* obligation to indemnify the other for losses due to tort claims. In many states a claim for indemnification states, essentially, that while the defendant-indemnitee may be legally or technically liable for the harm plaintiff has suffered, responsibility for the loss should be placed upon another, i.e., the indemnitor. In *Promaulayko v. Amtorg Trading Corp.*,[5] an indemnity claim following a suit brought by a worker exposed to asbestos shingles, the court confirmed that indemnity may not be recovered by a party that has itself been at fault, and stated the general rule: "A person who, without fault, has become subject to tort liability for the unauthorized or wrongful conduct of another, is entitled to indemnity from the other for expenditures properly made in the discharge of such liability."[6]

In addition to indemnification claims based upon actual judgments of liability, a party may also seek indemnification for expenditures undertaken in the good faith settlement of tort claims brought against it. A party whose relationship to the manufacture or marketing of an unreasonably dangerous product is itself vicarious may defend successfully against an indemnification claim brought by another vicariously liable respondent.[7]

17.2 CONTRIBUTION — GENERALLY

Even before liability is found, where one codefendant was engaged in the same actions for which liability is sought as to another defendant, either defendant may claim a right to contribution from the other for a share of any proved liability. The uniform law that has influenced many state laws governing contribution, in turn, provides that "where two or more persons become jointly or severally liable in tort for the same injury . . . even though judgment has not been recovered against all or any of them, [t]he right of contribution exists . . . in favor of a tortfeasor who has paid more than his pro rata share of the common liability[.]"[8]

The party seeking contribution from another must show that the other's negligent act or omission to act contributed substantially to the plaintiff's injury. Where the party from whom contribution is sought merely failed to discern the risks, and act to prevent them, contribution may not be appropriate. For example, in *Byrne v. SCM Corp.*,[9] a product vendor's contribution claim against an employer whose employee is injured by exposure to toxic fumes failed to hold the employer liable. In that suit, the court found that the manufacturer's warnings on the container of epoxy paint were insufficient to alert the painter to the dangers of using the paint without equipment, such as a mask or a respirator. Finding against the manufacturer, the court noted that no evidence showed that the employer should have known or should have investigated the need for using such equipment.

17.3 PARTIES TO AN INDEMNIFICATION CLAIM

The distributor or retailer of a hazardous product will often seek indemnification from the manufacturer or producer of the product. The rationale for such a claim is that, as a retailer or a distributor, defendant was only "passively" liable for the harm i.e., at most, as a downstream distributor the defendant failed to detect and remedy the hazard. The upstream manufacturer, the argument continues, was "actively" culpable for having designed, formulated, and initially marketed the product.

It can be seen that an "active-passive" standard for indemnification is an all or nothing proposition, in the sense that even a measurably culpable producer can avoid an indemnity obligation if it can prove that the distributor or retailer too was more than minimally incautious in selling the product. Accordingly, courts in some jurisdictions have abandoned the "active-passive" standard for indemnification, and rely instead upon comparative fault principles in allocating proportionate liability between multiple sellers. In

Dole v. Dow Chemical Co.,[10] where the underlying injury was the death of an employee who was overcome by exposure to the manufacturer's hazardous fumigant, the New York Court of Appeals held that the relative liabilities of the employer and the manufacturer should be apportioned according to their respective degrees of fault.

A government contractor producing products at the behest of the government procurement authority may, when later found liable for alleged defects in the product, seek indemnification from the government. A federal court recognized the potential for such a claim in *Johns-Manville Corp. v. United States,*[11] a suit brought by manufacturers of asbestos products arising from claims by or on behalf of World War II shipyard workers. Significantly, the manufacturers were unable to prevail in their claim due to the court's finding that to find the existence of an implied indemnification obligation would contravene the Anti-Deficiency Act prohibition against agency contracting beyond the limit of their appropriations.

Other asbestos-related indemnification claims brought by contractors against the government have been barred by operation workers' compensation law. In this setting, the injured employees or their estates, as former employees of the government, were precluded from a products liability/personal injury claim against the government. Under the law of some jurisdictions, this immunity from an employee personal injury claim operates likewise to bar a contractor claim against the government for indemnification, absent an express indemnification agreement.[12]

In other suits, asbestos contractor claims for indemnification from the government have been denied relief by application of other state and federal statutes. In *Lopez v. A.C. & S. Inc.*,[13] defendant's indemnification claim for settlement payments made to a pipe-fitter who had worked at the U.S. Naval Yard, the court agreed that the claim was not barred under the Federal Tort Claims Act,[14] but held nonetheless that the indemnity claim would fail because the settling plaintiff was a civil servant and not a harbor worker within the coverage of the federal Longshore and Harbor Worker's Compensation Act (LHWCA).[15] As additional grounds for denying the claim, the appeals court found that under the law of Washington state, the government, as an employer carrying appropriate workers' compensation insurance, would not be liable to indemnify others, i.e., the manufacturers, for damages arising from product-related injuries to a worker. In other suits brought by maritime workers to whom the LHWCA has been found to apply, indemnification claims against the government have been denied on the basis of the exclusion of federal employees from that act. In *Eagle-Picher Industries, Inc. v. United States,*[16] a federal appeals court held that the manufacturer's third-party indemnification claim, based upon the LHWCA, must fail. The court reasoned that "[a]ny other construction of the LHWCA would subject the United States to liability under a statute that from its inception has barred government employees from its coverage."

17.4 VOUCHING IN

Where an entity not a party to a suit will be liable to indemnify a defendant, the defendant may apply pressure upon the nonparty to join and defend the suit by the process known as "vouching in". The elements of a successful "vouching in" are (1) timely and sufficient notification and tendering of the defense to the nonparty and (2) the nonparty's knowledge of its ultimate liability should the vouching party be found liable. The significance of the vouching in mechanism is that if the entity who is properly notified of the claim fails to join the suit, a judgment against the party may be collected against the vouchee.

The benzene case of *Foremost-McKesson v. Allied Chemical Corp.*[17] is illustrative. In that suit a distributor of benzene sued by the representative of a woman who died of leukemia was found entitled to trial of its vouching-in action against the manufacturer of the chemical even where the manufacturer claimed that it could not know of its ultimate liability, arguing that the distributor failed to discover or warn adequately of the hazards posed by benzene.

NOTES

1. E.g., *Allied Corp. v. Frola,* 730 F. Supp. 626, 30 E.R.C. (BNA) 1954, 1965 (D. N.J. 1989) ("Indemnification is available under New Jersey Law in two situations: when a contract explicitly provides for indemnification or when a special legal relationship creates an implied right to indemnification.").
2. *Foremost-McKesson Corp. v. Allied Chemical Corp.,* 140 Ariz. 108, Prod. Liab. Rep. (CCH) ¶ 9884 (Ariz. App. 1983) (benzene).
3. 466 P.2d 722 (Cal. 1970).
4. *Id.*, 466 P.2d at 729.
5. 540 A.2d 893, Prod. Liab. Rep. (CCH) ¶ 11,828 (N.J. App. Div. 1988).
6. See Restatement, Restitution § 96 (1937).
7. *Promaulayko v. Amtorg Trading Corp.*, 540 A.2d 893, Prod. Liab. Rep. (CCH) ¶ 11,828 (N.J. Super. 1988) ("Where, as here, the distributor's liability is based on a common law failure to detect the defect in the product and this failure merely continued the defect created by the manufacturer, we perceive no valid reason to shift the liability of one distributor to another distributor through common law indemnification." *Id.* at 39,930.).
8. Uniform Contribution Among Tortfeasors Act § 1(a), (b).
9. 545 N.E.2d 105, Prod. Liab. Rep. (CCH) ¶ 12,170 (Ill. App. 1989).
10. 30 N.Y.2d 143, 282 N.E.2d 288 (1972).
11. 13 Ct. Cl. 72, Prod. Liab. Rep. (CCH) ¶ 11,320 (Cl. Ct. 1987).
12. *Lopez v. Johns-Manville,* 649 F. Supp. 149 (W.D. Wash. 1986).
13. 858 F.2d 712, Prod. Liab. Rep. (CCH) ¶ 12,058 (Fed. Cir. 1988).

14. 28 U.S.C. § 2764.
15. 33 U.S.C. §§ 903-904.
16. _____ F.2d _____, Prod. Liab. Rep. (CCH) ¶ 12,927 (1st Cir. 1988).
17. 140 Ariz. App. 108, 680 P.2d 818, Prod. Liab. Rep. (CCH) ¶ 9883 (Ariz. App. 1983).

CHAPTER 18

Damages

CONTENTS

Damages

18.1 COMPENSATORY DAMAGES

18.1.1 Generally

Where plaintiff is successful in proving that defendant's tortious conduct proximately caused toxic harm, the jury, or, in a suit tried without a jury, the trial judge, will award compensatory damages. These compensatory damages will be gauged as the amount of money it will take to place plaintiff in the financial position he would have been in absent defendant's wrongful conduct. It is immediately apparent that where plaintiff has suffered personal physical injury or disease from exposure to the toxin, an amount of money damages serves only as economic solace for an injury that is personal, emotional, and often debilitating. No pretense is maintained that compensatory damages can accurately repay a person who has suffered personal injury for their physical pain and suffering or their emotional distress. Similar incongruity exists where the available tort remedy is pecuniary and plaintiff's claims involve dignitary or emotional hardship, such as claims for loss of consortium, lost enjoyment of life, or rational fear that a prior toxic exposure increases the risk that plaintiff will contract disease in the future.

Other injury or damage from toxic harm is more suited to the logic of compensatory damages. Where a person has suffered personal physical injury, such claims may include past and future medical expenses, or cost of rehabilitation. Economic loss, such as lost income of a person or a business, may likewise be recovered in compensatory damages. For property damage or contamination occasioned by toxic exposure, plaintiff may seek money damages measured by the lost value of the property, be it measured by loss of resale value, or in the case of rental property, the lowering of its rental viability. Plaintiff may also seek the costs incurred, or costs that would necessarily be incurred, in restoring the property to its condition prior to toxic exposure.

18.1.2 Personal Injury — Generally

An individual suffering personal, physical injury or disease from exposure to a toxic harm may seek compensatory damages for past and future (1) physical pain and suffering; (2) emotional distress;[1] (3) medical and rehabilitation expenses; (4) lost income; and (5) in some jurisdictions, lost enjoyment of life. Depending upon the law of a particular state, a family member may have a claim for emotional distress or loss of consortium.

Claims for past and future medical expenses, or the cost of rehabilitation, often can be measured with accuracy. Even where not measurable with precision, rational estimates may be made of future medical, care provider, or rehabilitation costs.

In some instances of toxic harm, such as, for example, limited exposure to respirable asbestos, plaintiff's personal physical injury is minimal or incipient. It has been held that discrete cellular alterations caused by exposure to defendant's toxin may suffice for the definition of "injury" for statute of limitations purposes, but it remains a jury question as to whether it is compensable injury sufficient for liability in damages.[2]

18.1.3 Property Damage — Generally

Claims for toxic damage to property may be measured by calculation of the property value lost due to toxic damage or contamination. Loss of sales value, or for property put into service, loss of rental value, are typical measures for such damages. Depending upon the law of a particular jurisdiction, cost of remediation, cleanup, or repair may provide another means of calculating damages. Where such a rule is applied, remediation costs will not be allowed in damages to the extent these costs exceed the fair market value of the property.

In some jurisdictions, claimants alleging toxic tort damage to property alone must bring their cause of action within the conventional limitations period[3] and are held unable to avail themselves of any applicable discovery limitations rule.[4] Irrespective of a jurisdiction's adoption of a discovery interpretation to its limitations periods, plaintiff's claim for property damage will be deemed to accrue when the harm is capable of proof, rather than at some later time when the extent of harm may be measured to a certainty.[5]

18.1.4 Economic Loss Rule

Loss or damage other than personal physical injury or damage to property can take the form of lost wages, lost future income, or lost business opportunity. These are commonly called "economic" losses and are available to the tort plaintiff only upon demonstration of certain proof. Generally, where the toxic product or substance creates a risk of personal injury or property damage, plaintiff may, upon proper proof, recover economic loss proximately caused.

To be contrasted, where the mishap simply involves the failure of the product to perform as expected, and without personal injury or damage to property other than the product itself, the court may conclude that plaintiff's claim is not actually a claim in tort, but rather in warranty. Hence, for example, should a farmer purchase a specialized fungicide that, while yet in its original container, prematurely degrades and becomes useless, economic loss caused by any delay in his work would *not* involve a risk of personal injury or damage to property other than the product itself. As a consequence, any economic loss the farmer suffered, i.e., in the need to replace the fungicide or delays in work occasioned by the need to replace it, could only be recovered, if at all, by a claim in breach of express or implied warranty.

Accordingly, courts have concluded that plaintiff's claim for damage to the risk-producing product itself will not be obviated where plaintiff also demonstrates damage to other property.[6] The "damage to other property" test has not, however, been uniformly applied. For example, one court held that a pig breeder's suit against a feed seller claiming that nutrient and medication deficiencies in the feed caused the death of many pigs was one for economic damages and therefore sounded in warranty, and not tort.[7]

Toxic tort claims for property damage have been lodged by plaintiffs whose homes, in either their initial construction or later renovation, contain potentially hazardous building materials. *Fish v. Georgia-Pacific Corporation,*[8] a formaldehyde resin particleboard suit, involved a claim by plaintiff-homeowners for both personal injury to the husband and total loss of the home, on the basis of evidence that plaintiffs "had become permanently sensitized to formaldehyde, [and] that they could no longer live in the house[.]" The trial court instructed the jury that plaintiffs could recover for property damage if they showed that "dangerous defects in the particleboard rendered the home unsalable to other persons." The appeals court held that "unsalability" is not the proper test. Rather, the court stated, if liability was found, "the measure of damages on the property damage claim would be the cost of repairing the house, or, if this is not possible, the loss of value of the property because of the presence of particleboard in the home."

18.2 SPECIAL DAMAGES

18.2.1 Emotional Distress

The law in all jurisdictions permits the physically injured plaintiff to recover for mental, as well as physical, suffering.[9] Where defendant's conduct proximately caused pain and suffering, this doctrine permits compensatory damages for emotional distress. Emotional distress has been broadly described as including "fright, nervousness, grief, anxiety, worry, mortification, shock, humiliation, indignity, embarrassment, apprehension, terror or ordeal."[10] Courts in some jurisdictions permit recovery for emotional distress only upon plaintiff's demonstration of an underlying compensable harm. Thus plaintiff's exposure to asbestos has been held insufficient to support damages for emotional distress at an increased risk of contracting asbestos-related disease in the absence of plaintiff's proof of some compensable functional impairment.[11]

Several jurisdictions retain the "impact rule", and hold that damages for emotional distress "are recoverable only when accompanied by physical injury."[12] Significant to toxic tort matters, where exposure to toxins may result in cellular alteration that is barely perceptible, plaintiff's proof of exposure to X-ray radiation during pregnancy has sufficed to permit trial of a claim for emotional distress, even where there was no proved injury to the children born to that pregnancy.[13]

In some jurisdictions, courts confine plaintiff's claims for emotional distress damages to plaintiff's count in negligence, reasoning that the rationale for the remedy in strict products liability "was designed to compensate for economic — not emotional — damages."[14]

18.2.2 Increased Risk

An increasing number of toxic tort cases have permitted compensatory damages for emotional distress occasioned by an increased risk of suffering a toxic illness.[15] To prevail, plaintiff must prove "that future consequences of an injury are reasonably probable, and not just possible."[16]

Illustrative of plaintiff's claims is one suit in which plaintiffs alleged an increased risk of disease, including the risk of future cancer, from wellwater contamination by defendant's polyaromatic hydrocarbons. Plaintiffs' complaint claimed that plaintiffs had "undergone pain, mental anguish and emotional distress and will continue to do so, have experienced severe hysteria and fear, any or all of which will develop into a future psychiatric condition due, inter alia, to the acquired knowledge that they were unknowingly exposed to hazardous and harmful materials and can contract, as a result therefrom, cancer and other illnesses."[17] Such claims have also been allowed in actions where plaintiffs claim exposure to asbestos products[18] and to certain hydrocarbons.[19]

Plaintiff's claim for emotional distress for increased risk will fail in the absence of proof that he was exposed to the claimed hazardous agent. Thus, to use one example, plaintiff must do more than show that he might have been exposed to a needle previously used in connection with AIDS treatment.[20]

Courts declining to permit damages for increased risk state that the damages are "unquantified",[21] and that permitting such claims exposes the tort system to a vast number of claims for damages that may never occur.[22] A criticism added to that of a potential "flood of speculative suits"[23] is that award of such damages would decrease damages available to those who eventually did contract illness or disease, while providing a windfall to those who did not.[24]

In certain circumstances, plaintiffs may also recover for mental distress occasioned by their fear or apprehension of future illness or disease[25] and may include in their request for damages compensation for continued medical surveillance or monitoring to permit the early detection of future disease, should it occur.[26]

Where plaintiff claims emotional distress damages that are occasioned by a product defect that causes product malfunction, there can be no recovery until and unless the product actually fails. In the medical device context, a suit following a plaintiff heart valve recipient's discovery that the device with which he had been implanted had a defect that might cause it to fail, a Missouri appellate court held that plaintiff could state no claim for failure to warn, and consequent damages in emotional distress, absent the actual malfunction of the device in question.[27] In another heart valve suit, plaintiff was not permitted to recover damages for a manufacturer's omission to warn a heart valve recipient of risks associated with use of a *previous* design of the valve.[28]

18.2.3 Medical Monitoring

The toxic tort claimant may be awarded the reasonable costs of medical monitoring or surveillance.[29] A claim for medical monitoring costs differs from that for enhanced risk of contracting a disease in the future. A medical surveillance claim seeks to recover "the quantifiable costs of periodic medical examination necessary to detect the onset of physical harm," while the enhanced risk claim "seeks compensation for the anticipated harm itself, proportionally reduced to reflect the chance that it will not occur."[30]

18.2.4 Loss of Consortium

The toxic tort plaintiff's claim for loss of consortium turns upon proved "loss of conjugal fellowship and sexual relations," as well as "loss of love, companionship, society, sexual relations, and household services."[31] Plaintiffs have recovered damages for loss of consortium in suits arising from a spouse's injurious exposure to a variety of toxins.[32] The substantive law of some jurisdictions holds that plaintiff's claims for loss of consortium may sound in negligence, but not in strict tort liability.[33]

In *Anderson v. Eli Lilly & Co.*,[34] a New York appellate court stated that as consortium pertains to a couple's interest in the uninjured continuation of the marital relationship, a consortium claim could not be based upon the injury to one partner, sustained prior to marriage, but not detected until after the marriage.[35]

Some authority has limited the elements of consortium to love, companionship, society, sexual relations and solace, thereby leaving uncompensable a husband's loss of consortium claim that injuries caused by the wife's *in utero* exposure to DES left the couple unable to have natural children of the marriage.[36]

18.2.5 Loss of Enjoyment of Life

Plaintiffs have been permitted compensatory damages for loss of enjoyment of life where a toxic harm has significantly diminished the quality of the claimant's life. Factors bearing upon injury or loss of life's enjoyments might reasonably include impairments ranging from diminution of a plaintiff's mobility, loss of ability to operate motor vehicles or incapacity to appreciate the taste of food. Where a person is so severely injured as to be rendered unconscious, some courts have denied recovery for lost enjoyment of life, while others permit it. In one suit brought against an anaesthesiologist and a recovery room nurse on behalf of a patient who, following surgery, failed to regain consciousness for 1 year until his death, the court in *Eyoma v. Falco*[37] permitted recovery of damages for loss of enjoyment of life for that 1 year period of time. Other decisions, however, have denied this aspect of damages to the unconscious or comatose patient, reasoning that the injured person could not have any subjective appreciation of their loss or impairment.

In another circumstance, a potential parent exposed to a toxin might learn in a timely fashion that any children conceived would run a high risk of abnormal gestation. If, based on that risk, the couple decided to forbear having children, the frustrated father and mother alike might have a claim against the person or business responsible for the toxic exposure. In that setting, even if neither parent suffered diagnosable physical personal injury, their claim might be for lost enjoyment of life, or lost consortium, associated with the inability to bear children.

18.3 ECONOMIC LOSS

Claims that a person or a business has lost due to toxic harm can be proved with some accuracy. A claim for loss or diminution of future income can likewise be assessed, most frequently on the basis of the profit history of the person or the business. Greater problems of proof beset lost future income claims by or on behalf of younger persons, or persons not in the workforce at the time of the injury.

Where the damage to plaintiff's property results in a lessening of its income producing potential, plaintiff's damages will be measured by the difference between the economic benefits that were rationally predicted prior to the toxic harm, and the income producing potential, if any, following the harm.[38]

18.4 PUNITIVE DAMAGES

18.4.1 Generally

The purpose of punitive damages is not to compensate plaintiff for proved injury or loss, but rather to penalize or punish defendant and to deter defendant and others from engaging in such conduct in the future. The law of some states precludes recovery of punitive damages pursuant to plaintiff's claim in strict tort liability.[39] A government entity may seek an award of punitive damages in a common law action for public nuisance,[40] even where the conduct at issue may also be punishable as a crime.[41]

The general rule is that plaintiff's claim for punitive damages must be supported by evidence that defendant acted with conscious disregard of health or safety risks. Consistent with this approach, it has been held that 1932 Bureau of Mines documents regarding the danger of asbestos dust to manufacturing workers would not be admitted as evidence in suits brought decades later claiming seller knowledge of risks of asbestos exposure to ultimate users of such products.[42]

The standards for review of punitive damage awards, for excessiveness or inadequacy, were one issue before the Supreme Court in *Pacific Mutual Life Insurance Co. v. Haslip.*[43] There the Supreme Court found unobjectionable Alabama state court jury instructions inviting consideration of "(a) whether there is a reasonable relationship between the punitive damages award and the harm likely to result from defendant's conduct, as well as the harm that actually occurred; (b) the degree of reprehensibility of defendant's conduct, the duration of the conduct, the defendant's awareness, any concealment, and the existence and frequency of similar past conduct; (c) the profitability to the defendant of the wrongful conduct and the desirability of removing that profit and having the defendant also sustain a loss; (d) the 'financial position' of the defendant; (e) all the costs of litigation; (f) the imposition of criminal sanctions on the defendant for its conduct, these to be taken in mitigation; and (g) the existence of other civil awards against the defendant for the same conduct, these also to be taken in mitigation."[44]

18.4.2 Nature and Quality of Actor's Conduct

Sufficient "wantonness" was held shown in *Lakeman v. Otis Elevator Co.*,[45] a suit against the manufacturer of the active ingredient in a cleaning solvent, upon proof that an employee of the manufacturer had reviewed the warnings contained in the product's warning labels and had not objected, despite the labels' manifest inadequacy. Workers not directly working with the hazardous product may be classified as "bystanders", and face a correspondingly difficult task in proving the defendant's conduct was reckless or wanton as to *them.*[46]

18.4.3 Constitutional Limitations

Where a defendant's constitutional challenge to an award of punitive damages necessarily relies upon the excessive fines provision of the federal or state constitution, a reviewing court's task is complicated by the recognition that a substantial proportion of a punitive award may reflect the jury's wish to award damages on bases not associated with multiple punishment. In such circumstances, defendant's challenge may fail if it cannot show, by special jury interrogatory or otherwise, which proportion of the award might be tied to goals other than punishment and deterrence, such as, for example, reimbursement for losses too remote for ordinary compensatory damages, or compensation for inconvenience and attorneys' fees.[47]

NOTES

1. E.g., *Haas v. Celotex Corp.*, [not in F. Supp.], 1990 WL 42835 (E.D. Pa. 1990) (mesothelioma victim).
2. *Howell v. Celotex Corp.*, 904 F.2d 3 (3d Cir. 1990) (claimant's asbestos-related pleural thickening).
3. See discussion of statutes of limitation above, in Chapter 15.
4. E. g., *Corp. of Mercer Univ. v. United States Gypsum Co.*, 877 F.2d 35, Prod. Liab. Rep. (CCH) ¶ 12,202 (11th Cir. 1989) (on certified question to *Mercer Univ. v. National Gypsum Co.*, 258 Ga. 365, 368 S.E.2d 732 (1988)).
5. *Brower v. E. I. DuPont de Nemours & Co.*, 793 P.2d 345, Prod. Liab. Rep. (CCH) ¶ 12,640 (Idaho 1990) (farmer's claim for herbicide contamination of agricultural land).
6. *Kershaw County Bd. of Educ. v. U.S. Gypsum Co.*, 396 S.E.2d 369 (S.C. 1990) (school board's cause of action against manufacturer of asbestos products used in construction of educational properties).
7. E.g., *Drew v. United Producers and Consumers Cooperative*, 778 P.2d 1227 (Ariz. 1988) (en banc).
8. 779 F.2d 836, Prod. Liab. Rep. (CCH) ¶ 10,904 (2d Cir. 1985).
9. E.g., *Haas v. Celotex Corp.*, [not in F. Supp.], 1990 WL 42835 (E.D. Pa. 1990) ($300,000 for mesothelioma victim's physical pain and suffering).
10. *Rodriguez v. Bethlehem Steel Corp.*, 12 Cal. 3d 382, 115 Cal. Rptr. 765, 525 P.2d 669 (1974).
11. See *In re Hawaii Federal Asbestos Cases*, 734 F. Supp. 1563 (D. Hawaii 1990). In so ruling the court stated that in the absence of plaintiff's knowledge of compensable functional impairment, the generalized fear of cancer would not be reasonable.
12. E.g., *Rogers v. R. J. Reynolds Tobacco Co.*, 557 N.E.2d 1045, Prod. Liab. Rep. (CCH) ¶ 12,572 at 37,533 (Ind. App. 1990) (denying spouse's claim for emotional distress caused by illness suffered by husband).
13. *Jones v. Howard University, Inc.*, 589 A.2d 419 (D.C. App. 1991).
14. *Nasios v. Pennwalt Corp.*, ____ F. Supp. ____, Prod. Liab. Rep. (CCH) ¶ 12,479 at 37,046 (D. Md. 1990) (citing *Doe v. Miles Laboratories, Inc.*, 675 F. Supp. 1466 (D. Md. 1987)).
15. *In re Paoli Railroad Yard PCB Litigation*, 916 F.2d 829, 850 (3d Cir. 1990) (Pennsylvania law); *Sterling v. Velsicol Chemical Corp.*, 855 F.2d 1188, 1206 (6th Cir. 1988) (Tennessee law); Note, "Decreasing the Risks Inherent in Claims for Increased Risk of Future Disease," 43 U. Miami L. Rev. 1081 (1989).
16. *In re Paoli Railroad Yard PCB Litigation*, 916 F.2d 829, 850 (3d Cir. 1990) ((paraphrasing *Martin v. Johns-Manville Corp.*, 508 Pa. 154, 165 n.5, 494 A.2d 1088, 1094 n.5 (1985)).

17. *Stead v. F. E. Myers Co.*, _____ F. Supp. _____, (No. 89-169 (D. Vt. 1989).
18. E.g., *Griffin v. Keene Corp.*, [not in F. Supp.,.] 1990 WL 93292 (N.D. Ill. 1990) (denying manufacturer's motion in limine to prevent plaintiff's introduction of evidence regarding fear of contracting cancer or increased risk of contracting cancer due to exposure to defendant's asbestos products).
19. E.g., *Stead v. F. E. Myers Co.*, _____ F. Supp. _____, (No. 89-169) (D. Vt. 1989) (polyaromatic hydrocarbons, or PAHs).
20. In *Burk v. Sage Products, Inc.*, 747 F. Supp. 285 (E.D. Pa. 1990), a suit against a manufacturer of a medical device for the disposal of used medical syringes, plaintiff who claimed he was nevertheless stuck by a used needle could not prove that the needle had been used in AIDS-related treatment. Plaintiff's blood tests conducted one year after the incident were also negative for HIV antibodies.
21. *Mauro v. Owens-Corning Fiberglas Corp.*, 225 N.J. 196, 542 A.2d 16, Prod. Liab. Rep. (CCH) ¶ 11,876 (N.J. Super. 1988), citing "fundamental policy considerations dictating rejection of the unquantified risk claim." *Id.* at 34,139.
22. *Ayers v. Jackson Twp.*, 106 N.J. 557, 525 A.2d 287 (N.J. 1987) (plaintiff's enhanced risk of cancer claims against township for release of carcinogens in plaintiff's water supply "exposes the tort system, the public it serves, to the task of litigating vast numbers of claims for compensation based on threats of injuries that may never occur." *Id.* at 597.).
23. *Mauro v. Owens-Corning Fiberglas Corp.*, supra n. 21, Prod. Liab. Rep. (CCH) ¶ 11,876 at 34,139.
24. *Anderson v. W. R. Grace & Co.*, 628 F. Supp. 1219 (D. Mass. 1986) (" . . . the increased risk of future harm in this action is the inevitable inequity which would result if recovery were allowed. 'To award damages based on a mere mathematical probability would significantly undercompensate those who actually do develop cancer and would be a windfall to those who do not." *Id.* at 1232 (citation omitted).).
25. E.g., *Griffin v. Keene Corp.*, [not in F. Supp.], 1990 WL 93292 (N.D. Ill. 1990).
26. E.g., *Stead v. F. E. Myers Co.*, _____ F. Supp. _____, (No. 89-169) (D. Vt. 1989) (expert testimony permitted on issue of plaintiff's risk of future illness from exposure to polyaromatic hydrocarbons in wellwater, as well as reasonable requirement of continued medical monitoring for plaintiffs so exposed).
27. *Spuhl v. Shiley, Inc.*, 795 S.W.2d 573 (Mo. App. 1990).
28. See *Sill v. Shiley, Inc.*, 735 F. Supp. 337 (W.D. Mo. 1989) (valve implanted in plaintiff manufactured in 1986; manufacturer presented uncontroverted evidence that no valve failures occurred in valves manufactured after April, 1984).
29. E.g., *Ayers v. Township of Jackson*, 106 N.J. 557, 525 A.2d 287 (1987); *Merry v. Westinghouse Electric Corp.*, 684 F. Supp. 847 (M.D. Pa. 1988); *Burns v. Jaquays Mining Corp.*, 156 Ariz. 375, 752 P.2d 28 (App. 1988).
30. *In re Paoli Railroad Yard PCB Litigation*, 916 F.2d 829, 850 (2d Cir. 1990).
31. E.g., *Borer v. American Airlines, Inc.*, 19 Cal. 3d 441, 138 Cal. Rptr. 302, 563 P.2d 858 (1977).
32. E.g., *Haas v. Celotex Corp.*, [not in F. Supp.], 1990 WL 42835 (E.D. Pa. 1990) (asbestos exposure; $200,000 awarded to widow of mesothelioma victim).

33. *Nasios v. Pennwalt Corp.*, ____ F. Supp. ____, Prod. Liab. Rep. (CCH) ¶ 12,479 at 37,046 (D. Md. 1990). Citing *Doe v. Miles Laboratories*, 675 F. Supp. 1466 (D. Md. 1987), the court stated: "As to the loss of consortium claim, this Court noted [in *Doe*] that strict liability recovery was designed to compensate for economic — not emotional — damages." *Id.*
34. *Id.*, 158 A.D.2d 91, 557 N.Y.S.2d 981 (1990).
35. *Id.*, 158 A.D. 91, 557 N.Y.S.2d 981 (1990). In *Anderson* the court continued by stating its opinion that public policy would not be served by finding consortium liability based upon the surmise that the husband would not have married his wife had he known in advance of her incapacity to bear children due to her *in utero* exposure to DES.
36. *Anderson v. Eli Lilly & Co.*, 158 A.D.2d 91, 557 N.Y.S.2d 981 (1990).
37. 247 N.J. Super. 435, 589 A.2d 653 (N.J. App. Div. 1991).
38. *Brower v. E. I. Dupont de Nemours & Co.*, 792 P.2d 345, Prod. Liab. Rep. (CCH) ¶ 12,640 at 37,841 (1990) (describing plaintiff farmer's calculable loss arising from herbicide contamination, which precluded planting potato crops but not crops of wheat or barley, in these words: "[T]he proceeds of leasing land for potato crops over a four year period against the yield from wheat or barley;" held: plaintiff's claim barred by the statute of limitations).
39. *Nasios v. Pennwalt Corp.*, ____ F. Supp. ____, Prod. Liab. Rep. (CCH) ¶ 12,479 at 37,046 (D. Md. 1990), quoting *Doe v. Miles Laboratories*, 675 F. Supp. 1466, 1481 (D. Md. 1987) ("punitive damages are incompatible with a recovery based upon strict products liability.").
40. *United States v. Hooker Chemicals and Plastics Corp.*, 748 F. Supp. 67, 32 E.R.C. (BNA) 1203 (W.D.N.Y. 1990) (court declines to dismiss New York's punitive damages claim against chemical company in Love Canal litigation).
41. *Id.*, 32 E.R.C. (BNA) at 1207 ("In sum, it appears to the court that, under New York law, what the State seeks here is a civil penalty that is outside the purview of criminal law. . . . Contrary to Occidental Chemical Corporation's arguments, it appears that New York law has never deemed the creation or maintenance of a public nuisance to be punishable *only* as a crime and *only* by the statutory penalties enumerated in the Penal Law or its predecessor statutes.").
42. *In re Southern and Eastern Dist. Asbestos Litigation*, 730 F. Supp. 582 (S.D.N.Y. 1990).
43. ____ U.S. ____, 111 S. Ct. 1032 (1991).
44. *Id.*, 111 S. Ct. at 1045.
45. 930 F.2d 1547 (11th Cir. 1991).
46. *Eagle-Picher Industries, Inc. v. Balbos*, ____ A.2d ____, 1990 WL 124699 (Md. App. 1990) (worker who dies of mesothelioma but who did not actually work with defendant's asbestos products failed to prove manufacturer's reckless or wanton conduct sufficient to support an award of punitive damages).
47. *Celotex Corp. v. Tate*, ____ S.W.2d ____, 1990 WL 127301 (Tex. App. 1990).

CHAPTER 19

Representative Products by Class

CONTENTS

Representative Products By Class

19.1 GENERALLY

The types of products that have and will continue to give rise to toxic tort claims are seemingly inexhaustible and range from polychlorinated biphenols (PCBs) to dietary supplements.[1] This chapter undertakes the overview of specific and representative products that have been the subject of toxic tort claims. While it would be impossible to catalogue all such products, a summary of representative claims, by product or use category, benefits the reader seeking a sense of the toxic tort implications of the manufacture and marketing of certain classes of products.

Many of the references to follow involve suits discussed earlier. At the end of the chapter, particularized discussion is devoted to two residential toxic hazards, radon and the termiticide chlordane.

19.2 INDUSTRIAL OR MANUFACTURING SUBSTANCES

In a toxic tort suit deriving from a worker's fatal chemical pneumonia following exposure to the vapors and residue of the cleaning and degreasing solution 1-1-1-Trichloroethane,[2] the trial and appellate court held that the manufacturer's warning concerning inhalation of vapors was adequate to advise users to likewise avoid inhalation of the mist created by its pump spray application.[3] Manufacturers and distributors of naptha who provided adequate warnings to their purchaser were found not liable for failure to provide adequate warnings to an employee of the purchaser who was injured after taking some of the chemical home and using it in her washing machine to remove glue stains from a work apron.[4]

Toxic tort claims against the manufacturer of cleaning solvents have alleged injuries such as from Raynaud's phenomenon, allegedly caused by the solvent's defatting and evaporative effects.[5] In *Peterson v. Sealed Air Corp.*,[6] a federal court of appeals reviewed service of process issues involving the manufacturer of methyl isocyanate, as well as the amenability to suit of the manufacturer's subsidiary, formed to hold and license the product trademark.

A seller of cobalt to a manufacturer of carbide tools was found not liable for injuries sustained by a worker exposed to cobalt dust where the seller argued successfully that the tool manufacturer was a "sophisticated user" to whom the supplier owed no duty to warn.[7] Foundry workers have prosecuted claims against suppliers of silica sand, claiming that the suppliers omitted to provide adequate warnings of the risks of prolonged inhalation of silica dust.[8]

Workers exposed to zinc products defendant manufacturers sold to plaintiffs' employer have brought claims that their preexisting multiple sclerosis was accelerated or worsened by the exposure.[9]

19.3 ASBESTOS PRODUCTS

Tort claims without seeming limitation continue to be lodged against manufacturers and other sellers of products containing respirable asbestos fibers. A significant proportion of such claims allege that the manufacturer or contractor failed to provide asbestos workers, or jobsite bystanders, with adequate warnings of the hazards or equipment necessary to reduce or eliminate the hazard.[10] A claim may sound not only in negligence or strict products liability,[11] but in warranty as well.[12]

The asbestos products at issue in these claims range from asbestos cloth employed in submarine building,[13] to cement products containing asbestos,[14] to asbestos-impregnated tile products.[15] Some courts have held that the rule precluding recovery of purely economic loss will not relegate plaintiff's claims to the rules of warranty where the manufacturer's asbestos-containing product caused damage to other property.[16]

19.4 CHEMICAL BYPRODUCTS OR WASTE

Disposal of chemical waste harming nearby property is often redressable by a toxic-tort suit.[17] For example, contamination emanating from defendant's maintenance of a waste site may interfere with a nearby land occupier's use and enjoyment of their property, creating a claim in nuisance.[18]

A variety of toxic tort actions involve claimed toxic contamination by PCBs.[19] Gaseous effluent from smokestacks or elsewhere may create a claim in trespass or nuisance.[20] Suits have been based upon airborne waste containing toxins of a metallic or other origin,[21] as well as airborne sulfur dioxide waste from a defendant's coal burning electric generating plant.[22]

19.5 PETROCHEMICALS AND PETROLEUM-BASED PRODUCTS

Actions in nuisance or trespass have been maintained against defendants engaged in the manufacturing or distribution of petroleum-based products from asphalt[23] to fuel oil.[24] Multiple claims in negligence, warranty, and strict libility have derived from wellwater contamination by polyaromatic hydrocarbons.[25]

A limited number of actions brought by roofing or highway construction workers and others claiming respiratory or skin-related injury from exposure to asphalt products have failed to prove either the unreasonably dangerous nature of the product or its causal contribution to plaintiff's illness.[26]

19.6 RADIATION

Plaintiffs who have undergone radiation therapy have pursued claims that the health care provider failed to properly shield other body areas from radiation.[27] Actions have also been filed against a manufacturer whose predecessor produced Thorotrast, a radioactive contrast dye used in diagnostic X-rays,[28] and against a landlord by a premises occupant claiming injurious exposure to X-rays incident to operations in a nearby office in the same building.[29]

Plaintiffs who were pregnant at the time of exposure to X-rays have been permitted to pursue claims for emotional distress even where the pregnancy resulted in the birth of healthy offspring.[30] One plaintiff sued his employer alleging injury from exposure to electromagnetic pulse radiation, achieving a substantial settlement in which the employer also agreed to provide 10 annual medical examinations to over 700 current and former employees.[31]

19.7 PESTICIDES, FUNGICIDES, AND RODENTICIDES

Toxic tort claims have been lodged against manufacturers of pesticides where the purchaser claimed that use of the manufacturer's herbicide prevented the timely planting of plaintiff's regular crop rotation.[32] One suit, brought by a golf course employee, alleged that a fungicide manufacturer's failure to provide adequate warnings about the risk of mercury poisoning from exposure to its product was the proximate cause of his injury.[33] A herbicide seller's erroneous claims that its product will not harm a strawberry grower's produce may create a claim in breach of express warranty.[34] Suits have also been brought by homeowners against manufacturers and distributors of diverse fumigating agents.[35]

19.8 TOBACCO PRODUCTS

Several decisions have found that the Cigarette Labeling and Advertising Act[36] impliedly preempts state law claims that tobacco manufacturers failed to provide adequate warnings of risks associated with the use of tobacco products. Other decisions have found as a matter of law that cigarettes are not "unreasonably dangerous".[37] Nevertheless, there is authority suggesting that findings of implied preemption do not preclude claims that a tobacco product is unreasonably dangerous under state law.[38]

19.9 MEDICAL, DENTAL, AND BIOLOGICAL PRODUCTS

In many jurisdictions, medical devices, dental products, and biological products have been accorded so-called "comment k" status under Restatement, Second, Torts § 402A. Comment k provides that manufacturers and sellers of certain products deemed "unavoidably unsafe" should not be liable in strict tort liability if the product was manufactured and sold with adequate warnings and instructions for use.[39]

The rationale for applying comment k analysis to these products tracks that stated for application of comment k to pharamaceuticals. The policy reasons for precluding strict tort liability for providers of prescription medical products have been summarized as including (1) an apprehension that strict tort liability will stifle research, development, and distribution of beneficial drugs; (2) imposition of strict tort liability would raise manufacturer's insurance rates to a crippling level; and (3) imposing tort standards of care extending beyond the rigorous premarket requirements of the Food and Drug Administration would not serve the public interest.[40]

Producers of blood products used in the treatment of hemophiliacs have also been made defendants in actions brought by users claiming that the manufacturer's blood product transmitted the AIDS virus or associated antibodies.[41] For injuries by transfusion prior to development of an efficacious test for detecting the AIDS virus in blood supplies, the majority of courts have held that blood products should be treated as "unavoidably unsafe" within the meaning of comment k to Restatement, Second, Torts § 402A.[42]

In addition, in most jurisdictions, blood-based biological products are immunized from strict tort liability by statute. These states have enacted "blood shield" statutes that materially limit a toxic tort claimant's cause of action for contaminated blood.[43] Different approaches are adopted between and among such statutes, with many declaring a transfusion to be a service, rather than a sale, and thus beyond the reach of strict products liability or warranty liability.[44] Statutes in other jurisdictions require that plaintiffs prove that the seller or distributor of the biological product was negligent, while others preclude tort liability altogether where the defect or contamination cannot be detected. Neither "blood shield" statutes nor comment k should, however, preclude tort recovery against the provider of contaminated blood where the seller failed to utilize then-existing screening measures or safeguards.[45]

While a medical device manufacturer's duty to warn of the risks associated with the use of its product extends only to the medical community, plaintiff may yet prove liability upon a showing that the warnings given inadequately apprised physicians of the product's risks.[46] Regarding toxic constituent elements in dental products used in fillings and oral surgery, in *Dorton v. Landmark Dental Care*,[47] an Alabama court overturned a jury verdict absolving a dentist sued for using Sargenti paste, which contains toxic materials, in a patient's root canal procedure.

19.10 RESIDENTIAL HAZARDS

19.10.1 Formaldehyde

Plaintiffs have raised triable issues of fact as to whether urea-formaldehyde foam insulation is causally connected with health problems of plaintiff homeowners.[48] For example, in *Shorter v. Champion Home Builders Co.*,[49] the homeowners claimed that the particleboard used by the defendant homebuilder emitted "dangerous levels of formaldehyde" that caused "various medical problems[,] including . . . eye irritation, throat irritation, headaches, lethargy, runny noses, congestions, inflamed nasal passages, wheezing, asthmatic symptoms, flu-like symptoms, and emotional harm." Other formaldehyde-related suits have been brought by plaintiffs claiming harmful exposure to vapors emitted from furniture.[50] To date, use of products containing formaldehyde in medical or dental applications has not supported plaintiff's malpractice action against dental care providers.[51]

19.10.2 Chlordane

Chlordane is a chlorinated cyclodiene insecticide. Its use as an insecticide has been recognized since 1945, and it has been produced in the U.S. since 1947. Chlordane and heptachlor are registered for use as termiticides under the Federal Insecticide, Fungicide, and Rodenticide Act (FIFRA).[52] They are considered particularly effective because migration is minimal, and the products can remain in soil for as long as 30 years. In 1982 the Environmental Protection Agency (EPA), acting under FIFRA, undertook investigation of the environmental effects of chlordane. On August 11, 1987, the EPA and Velsicol, manufacturer of chlordane, entered into an agreement under which Velsicol agreed to cease all sales and distribution of chlordane.

Much of the case law dealing with chlordane raises evidentiary questions, including the admissibility and inadmissibility of expert testimony amd studies. In one action against an exterminator, *Villari v. Terminix International*,[53] plaintiffs were allowed to introduce the results of animal studies to prove carcinogenicity. In the same case, an expert was allowed to testify to the toxicity of chlordane despite the fact that evidence that the substance had been withdrawn from the market was found to be inadmissible.

Regarding plaintiffs' claims that exposure to chlordane increased their risk of contracting disease in the future, plaintiffs must put forth expert testimony that quantifies their risk of disease. Some decisions hold that plaintiffs must show that they would "most probably" suffer from the diseases they claim they may contract due to their exposure.[54] In many states, present damage recovery for an increased risk of contracting a disease in the future is not recoverable without evidence of present injury.

Another issue frequently litigated with respect to chlordane is the admissibility of evidence of subsequent remedial measures in strict products liability cases.[55] In one termiticide case brought against an exterminating company, *Dine v. Western Exterminating Company*,[56] plaintiffs sought to introduce evidence of an 1987 agreement between Velsicol and the EPA to limit the registered uses of the chemicals. Defendant objected to its admissibility because they anticipated it would be used as evidence that the particular insecticide at issue was a defective product, and that plaintiffs intended to use the 1987 agreement as evidence that Velsicol was aware of the hazards of chlordane and heptachlor. While holding that evidence of the agreement between the EPA and Velsicol was inadmissible under Federal Rule of Evidence 407, the court further held that the evidence was not probative of the toxicity or carcinogenicity of chlordane and heptachlor and a jury could not have drawn such an inference.

Likewise, in *Villari*,[57] plaintiff homeowners were precluded from introducing evidence that the defendant exterminator had ceased using chlordane and similar chemicals as part of their proof that these chemicals were hazardous. The court found that the exterminator's action was a subsequent remedial measure and was therefore inadmissible, and held further that evidence that termiticides containing chlordane and other chemicals had been taken off the market was also inadmissible to show that the termiticide was hazardous.[58]

19.10.3 Radon

Homeowners discovering that their home is radon contaminated may have a toxic tort claim. Most frequently, the claim will be brought against the architect of the home, the builder, or both parties. In such a suit, the plaintiff may seek relief on the theories of strict liability, the implied warranty of habitability, negligence, or fraudulent concealment.

To establish a cause of action for strict products liability, the plaintiff must demonstrate that: (1) radon contamination poses an unreasonable risk of personal injury; (2) as contaminated, the home is in a defective condition that is unreasonably dangerous to its inhabitants; (3) the defect was present at the time of sale; and (4) the manufacturer failed to warn of any danger associated with the dwelling.

For the purposes of a toxic tort claim arising from radon contamination, the defect in the dwelling could be inadequate sealage, which allows radon gas to enter the dwelling; inadequate ventilation, which permits the radon gas to remain in the dwelling; or contaminated building products, which provide the source of the radon gas.

Where recognized, the implied warranty of habitability is interpreted as meaning that at the time a vendor delivers a home for sale, he warrants that the structure is (1) free from faulty materials; (2) constructed according to sound engineering standards; (3) constructed in a workman-like manner; and (4) fit for habitation.

In a negligence claim arising from radon contamination, the plaintiff must establish that the builder knew or should have known of the possibility of the contamination. A homeowner may also claim that the seller fraudulently concealed the hazardous contamination. Such a claim can be pursued against either the builder of the radon-contaminated home or against the manufacturer of a contaminated product used in its construction, if appropriate. For example, in *Wayne v. Tennessee Valley Authority,*[59] plaintiffs owned a home built with concrete blocks into which was incorporated phosphate slag. The phosphate slag emitted radon gas. They brought an action in negligence, implied warranty, and strict liability against the producer of the phosphate slag. With respect to plaintiff's claim that the producer of the phosphate slag fraudulently withheld knowledge of its radioactivity, the federal appellate court found for the defendants. The court reasoned that as the Tennessee Valley Authority (TVA) "did not know or have reason to know of the danger of radon gas in 1969, when the blocks at issue were sold, and did not know or have reason to know of such a danger until near the end of the 1970's, when TVA stopped the sale of phosphate slag and told the public why it was doing so, producer did not fraudulently withhold knowledge of hazardous radioactive nature of phosphate slag and slag-filled concrete blocks."[60]

Also illustrative is the Colorado decision in *Schnell v. Gustavson,*[61] a home purchasers' fraud suit against the prior owners of the home and the corporation that sold the home to them. Evidence in that suit showed that the previous owners of the home were aware that there were mill tailings beneath their home, as was the corporation, but that neither party informed the plaintiffs. The court held that the lack of contractual privity between the prior homeowners and the current homeowners did not relieve the prior homeowners of their duty to disclose this latent defect.

In *Brafford v. Susquehanna,*[62] plaintiffs were family members whose house was built on the site of a uranium-milling facility and as a result were exposed to radiation at levels which far exceeded government regulatory standards. They brought an action in, among other things, negligence, failure to warn, and strict liability against the last owners of the mill, Susquehanna Western, Inc., and alleged that mill tailings which emit radon gas were placed around the foundation of their home. There the federal appeals court held that plaintiffs could not state a claim for strict liability on the basis of increased risk of cancer. In order to recover future damages, plaintiffs must demonstrate that they have sustained a present injury as well. Significantly, the court also held that plaintiffs' claim of irreparable chromosomal damage was sufficient to defeat defendant's motion for partial summary judgment, although the issue whether such damage constituted present injury was not decided.

NOTES

1. E.g., *Duryea v. Showa Demko, K. K.*, 1990 Toxics Law Rptr. 788 (Nov. 21, 1990) (claim against Japanese manufacturer and domestic distributors of L-tryptophan for neurological injuries allegedly sustained by women using the dietary supplement).
2. *Pruitt v. P. P. G. Industries, Inc.*, ____ F.2d ____, Prod. Liab. Rep. (CCH) ¶ 12,397 (11th Cir. 1990).
3. The manufacturer's duty to warn is discussed in Chapter 9.
4. *Whitehead v. Dycho Co.*, ____ S.E.2d ____, Prod. Liab. Rep. (CCH) ¶ 12,215 (Tenn. 1989).
5. *Bond v. Octagon Process, Inc.*, 745 F. Supp. 710 (M.D. Ga. 1990), *aff'd*, 926 F.2d 1573 (11th Cir. 1991), *cert. denied*, ____ U.S. ____, 111 S. Ct. 2855 (1991).
6. ____ F.2d ____, 1990 WL 51441 (7th Cir. 1990).
7. *Tasca v. GTE Products Corp.*, Prod. Liab. Rep. (CCH) ¶ 12,122 (Mich. App. 1988).
8. *Smith v. Walter C. Best, Inc.*, ____ F. Supp. ____, 1991 WL 16835 (W.D. Pa. 1991). In that suit, the Pennsylvania trial court, applying Ohio law, found no seller duty to warn the "knowledgeable purchaser" of its product. For an analysis of these and comparable "duty to warn" issues, see Chapter 9.
9. *Michael v. Ametelco, Inc.*, ____ N.Y.S.2d ____, 1991 WL 60065 (N.Y. Sup. 1991) (held: claim of aggravation of preexisting disease was not a separate or distinct injury or disease sufficient to trigger a new statute of limitations period).
10. E.g., *Borel v. Fibreboard Paper Products Corp.*, 493 F.2d 1076 (5th Cir. 1973), rehearing denied *en banc; Graham v. Pittsburgh Corning Corp.*, 593 A.2d 567 (Del. Super. 1990).
11. E.g., *Dillon v. Fibreboard Corp.*, 919 F.2d 1488, Prod. Liab. Rep. (CCH) ¶ 12,706 (10th Cir. 1990).
12. *Cf.*, *Caccavale v. Raymark Indus., Inc.*, 404 Mass. 93, Prod. Liab. Rep. (CCH) ¶ 12,108 (Mass. 1989).
13. *Champagne v. Raybestos-Manhattan, Inc.*, 212 Conn. 509, Prod. Liab. Rep. (CCH) ¶ 12,247 (Conn. 1989).
14. *Foster v. Fibreboard Corp.*, 55 Wash. App. 545, Prod. Liab. Rep. (CCH) ¶ 12,280 (Wash. App. 1989).
15. *Berkow v. Celotex Corp.*, 722 F. Supp. 1547 (N.D. Ill. 1989).
16. E.g., *Kershaw County Bd. of Educ. v. U.S. Gypsum Co.*, 396 S.E.2d 369 (S.C. 1990) (school board's action against asbestos product manufacturer for damage caused to school property by construction with asbestos-containing products and consequent required removal).
17. E.g., *Vicksburg Chemical Co. v. Thorness*, 355 So.2d 299 (Miss. 1978).
18. E.g., *Cooper v. Armstrong Rubber Co.*, 29 E.R.C. (BNA) 1102 (S.D. Miss. 1989).
19. E.g., *United States v. Carolina Transformer Co., Inc.*, ____ F.2d ____, 31 E.R.C. (BNA) (E.D.N.C. 1989).
20. E.g., *Morgan v. High Penn Oil Co.*, 238 N.C. 185, 77 S.E.2d 682 (1953).

21. E.g., *Bradley v. American Smelting and Refining Co.*, 104 Wash. 2d 677, 709 P.2d 782 (Wash. 1985) (action in trespass arising from deposit of microscopic particles of arsenic and cadmium).
22. See *Jost v. Dairyland Power Cooperative*, 45 Wis.2d 164, 172 N.W.2d 647 (Wis. 1969).
23. E.g., *Rushing v. Hooper-McDonald, Inc.*, 293 Ala. 56, 300 So.2d 94 (Ala. 1974) (asphalt contamination).
24. *Serota v. M. & M. Utilities, Inc.*, 55 Misc. 2d 286, 285 N.Y.S.2d 121 (1967) (action in trespass arising from mistaken delivery of fuel oil).
25. *Stead v. F. E. Myers, Co.*, ____ F. Supp. ____, (No. 890-169) (D. Vt. 1990).
26. *Goode v. Tamko Asphalt Products, Inc.*, ____ S.E.2d ____, Prod. Liab. Rep. (CCH) ¶ 12,318 (Tenn. 1989) (claim of vitiligo, causing change in skin pigmentation).
27. E.g., *Ellingwood v. Stevens*, 564 So.2d 932 (Ala. 1990) (approving admissibility of testimony of plaintiff's expert, a physicist, on issue of proper shielding of plaintiff's spinal cord during radiation therapy).
28. *Simmers v. American Cyanamid Corp.*, ____ A.2d ____, 1990 WL 78091 (Pa. Super. 1990).
29. *Star v. Berridge*, 568 N.Y.S.2d 904, 571 N.E.2d 74 (N.Y. 1991) (held: landlord had no duty in tort).
30. *Jones v. Howard University, Inc.*, 589 A.2d 419 (D.C. App. 1991).
31. *Strom v. Boeing Co.*, ____ P.2d ____, (Wash. 1990), 1990 Toxics Law Rptr. 404 (Aug. 22, 1990) (No. 88-2-10752-1) (settlement approved 8-15-90).
32. *Brower v. E. I. DuPont de Nemours & Co.*, Prod. Liab. Rep. (CCH) ¶ 12,640 (Idaho 1990) (farmer's claim that herbicide unduly delayed planting of crops other than wheat or barley; held: plaintiff claim barred by statute of limitations).
33. *Fitzgerald v. Mallinckodt, Inc.*, 681 F. Supp. 404, Prod. Liab. Rep. (CCH) ¶ 11,841 (E.D. Mich. 1987).
34. *Venie v. South Central Enterprises*, 401 S.W.2d 495, Prod. Liab. Rep. (CCH) ¶ 5531 (Mo. App. 1966) (hormone herbicide 2,4,5-T).
35. E.g., *Begley v. Truly Nolan Exterminating, Inc.*, ____ So.2d ____, 1991 WL 11731 (Fla. App. 1991), in which a homeowner's strict liability claim was held preempted by Federal Insecticide, Fungicide and Rodenticide Act (FIFRA). A later Supreme Court decision regarding FIFRA and preemption suggests that the *Begley* holding was incorrect. See the discussion of statutory preemption of tort claims in Chapter 14.
36. 15 U.S.C. §§ 1331–1341.
37. *Gilboy v. American Tobacco Co.*, ____ La. App. ____, 1990 Toxics Law Rptr. (BNA) 821 (Nov. 28, 1990) ("We determine as a matter of law that cigarettes are products that are not unreasonably dangerous per se.").
38. *Pennington v. Vistron Corp.*, 876 F.2d 414 (5th Cir. 1989).
39. Restatement, Second, Torts § 402A, comment k, provides, in pertinent part: "*Unavoidably unsafe products*: There are some products which, in the state of human knowledge, are quite incapable of being made safe for their intended and ordinary use. These are especially common in the field of drugs. . . . Such a product, properly prepared and accompanied by proper directions and warnings, is not defective, nor is it *unreasonably* dangerous."

40. See *Pollard v. Ashby,* ____ S.W.2d ____, Prod. Liab. Rep. (CCH) ¶ 12,327 (Mo. App. 1990) ("Holding drug manufacturers to a strict liability standard might deter research and development of new drugs or the distribution of available drugs that would prove beneficial. The cost of insurance will diminish the availability and increase the price of pharmaceuticals. Furthermore, withholding a drug from the market, beyond the delay required by the Food and Drug Administration, until scientific skill and knowledge advanced so that additional dangerous side effects might be revealed, would not serve the public interest." *Id.*, ¶ 12,327 at 36,345–36,346.).
41. E.g., *Poole v. Alpha Therapeutic Corp.*, 696 F. Supp. 351, Prod. Liab. Rep. (CCH) ¶ 12,047 (N.D. Ill. 1988). In Poole, plaintiff brought his action on behalf of a hemophiliac who died of AIDS following use of defendant's antihemophilic blood product known as factor VIII.
42. Restatement (Second) Torts 402A, comment k. E.g., *Miles Laboratories v. Doe,* ____ A.2d ____, Prod. Liab. Rep. (CCH) ¶ 12,126 at 35,330-31 (Md. App. 1989); *Kozup v. Georgetown University,* 663 F. Supp. 1048, 1059, Prod. Liab. Rep. (CCH) ¶ 11,943 (D.D.C. 1987), aff'd, 851 F.2d 437 (D.C. Cir. 1988) (the language of comment k "is in no way limited to hospitals as providers, and applies equally cogently to the ARC in the context of AIDS in blood in 1983").
43. See generally *Roberts v. Suburban Hospital,* 73 Md. App. 1, 532 A.2d 1081, Prod. Liab. Rep. (CCH) ¶ 11,583 (1988) (listing 48 states having enacted such statutes).
44. E.g., Chapter 259 of the Acts of 1986, Maryland Code, Section 18-402, which provides: "A legally authorized person who obtains, processes, stores, distributes, or uses whole blood or any substance derived from blood for injection or transfusion into an individual for any purpose is performing a service and is not subject to: (1) strict liability in tort; (2) the implied warranty of merchantability; or (3) the implied warranty of fitness."
45. E.g., *Doe v. University Hosp. of N.Y. Univ. Medical Center,* 149 Misc.2d 756, 5561 N.Y.S.2d 326 (N.Y. Sup. 1990) (claim of negligence liability arising from transfusion patient's infection with the AIDS virus).
46. E.g., *Amore v. G. D. Searle & Co., Inc.*, ____ F. Supp. ____, 1990 WL 156813 (S.D. Fla. 1990) (denying manufacturer's motion for summary judgment in suit brought by user of Copper-7 intrauterine contraceptive device, claiming that use of the product caused chronic pelvic inflammatory disease).
47. 577 So. 2d 425 (Ala. 1991).
48. *Zimprich v. Stratford Homes, Inc.*, ____ N.W.2d ____, 1990 WL 35792 (Minn. App. 1990).
49. ____ F. Supp. ____, 1991 WL 212776 (N.D. Ohio 1991).
50. E.g., *Allen v. Roddis Lumber and Veneer Co.*, 796 S.W.2d 758 (Tex. App. 1991), error denied.
51. E.g., *Koslowski v. Sanchez,* 563 So.2d 937 (La. App. 1990) (root canal filler use of N-2 paste containing 6.5% paraformaldehyde; plaintiff's cause of action for paresthesia).
52. 7 U.S.C § 136.
53. 692 F. Supp. 568 (E.D. Pa. 1988).
54. *Rabb v. Orkin Exterminating Co.*, 677 F. Supp. 424 (D.S.C. 1987).

55. See discussion in Chapter 11 pertaining to the application of Fed. R. Ev. 407 and comparable state provisions to introduction of postincident remedial measures.
56. [Not in F. Supp], 1988 WL 28241 (D.D.C. 1988).
57. Supra, 692 F. Supp. 568 (E.D. Pa. 1988).
58. See also *Rabb v. Orkin Exterminating Co.*, 677 F. Supp 424 (D.S.C. 1987) (evidence that exterminator discontinued use of termiticide and that manufacturer temporarily took termiticide off market inadmissible especially when product was still approved by EPA and state law).
59. 730 F.2d 392 (5th Cir. 1984).
60. *Id.* at 396.
61. 638 P.2d 850 (Colo. App. 1981).
62. 586 F. Supp. 14 (D. Colo. 1984).

CHAPTER 20

Insurance

CONTENTS

Insurance

20.1 THE COMPREHENSIVE GENERAL LIABILITY POLICY

20.1.1 Coverage and Exclusions — Generally

In general, a policy's failure to provide expressly for a pertinent liability coverage can be interpreted as an understanding by the carrier and the insured that risks not mentioned are not covered.[1] The insurance industry's primary form of commercial insurance coverage is the comprehensive general liability (CGL) policy. Originally, under a CGL policy an insurance company would pay on the behalf of the insured all sums the insured became legally obligated to pay as a result of an ''accident''. Beginning in 1966 the insurance industry changed the CGL policy to read that the insurance company would be liable for the insured's legal obligations which were caused by an occurrence. The term occurrence was defined to include accidents which were neither expected nor intended from the standpoint of the insured. One purpose of the change was to eliminate the ongoing litigation reaching variable interpretations, as to what did, and what did not, constitute an insured ''accident''.

Specifically, the CGL policy language governing a carrier's obligation to defend or indemnify a liability claim lodged against an insured reads as follows:

> [The insurer] will pay on behalf of the insured all sums which [the insured] is legally obligated to pay as damages because of bodily injury . . . to which this insurance applies, caused by an occurrence, and the [carrier] shall have the right and duty to defend any suit against the insured seeking damages on account of such bodily injury . . . even if any of the allegations of the suit are deemed groundless, false or fraudulent . . . .[2]

The policy defines ''bodily injury'' as ''bodily injury, sickness or disease sustained by any person.'' ''Occurrence'' is defined as ''an accident, including injurious exposure to conditions, which results, during the policy period, in bodily injury.''

20.1.2 The Duty to Defend

In *Ogden Corporation v. The Travelers Indemnity Co.*,[3] the federal appellate court explained that the duty to defend is ''derived from the allegations of the complaint and the terms of the policy.''[4] So long as the allegations of the complaint even potentially require coverage the insurer is obligated to defend or else has the burden of proving otherwise.[5] There is no duty to defend when the insurer establishes that there is no possibility that the allegations fall outside the exclusion. One court explained this rule in *State of New York v. Amro,*[6] stating that the insurer has a duty to defend when the pleadings allege a potentially covered occurrence even when facts outside the pleadings indicate the claim may be meritless or outside the scope of coverage.[7]

In interpreting obligations of carriers to defend or indemnify insured whose products are claimed to have caused long-latency harms, most suits have required courts to allocate responsibility between and among multiple carriers on the risk for any particular year or other time period. In such settings, most courts have prorated carrier responsibility according to the injuries sustained, and thus insured actual or potential liability during the years the particular carrier's policy was in effect.[8]

A growing number of decisions have adopted the rationale of *Keene Corp. v. Insurance Corp. of North America,*[9] which held that duties under the CGL policy are triggered from the claimant's initial exposure to asbestos through and including the manifestation of disease symptoms.[10] These courts have interpreted the disease-incapacitation continuum of asbestos-related injury as comprised of a series of discrete cellular injuries to the lungs. Under a *Keene* approach, the financial obligations of multiple carriers found to be on the risk may be prorated according to the length of time its policies were in effect during the entire asbestos disease process.[11]

20.1.3 The Meaning of "Occurrence"

For long latency harms resulting in property damage but not physical injury, many courts have concluded that coverage is triggered when actual property damage first occurs. Thus, where defendant's mortar additive used in building construction was claimed to cause structural damage in a series of buildings in which it was used, it was held that each building constructed with the claimed defective product was a separate "occurrence" within the meaning of the insured's general liability coverage.[12]

20.2 THE POLLUTION EXCLUSION

20.2.1 Generally

The "pollution exclusion" clause arose during a period where people were becoming increasingly alert to environmental hazards and their consequences. During this period, the Environmental Protection Agency (EPA) published estimates of the seemingly incalculable cost of cleaning up the country's worst hazardous-waste dump sites. Accordingly, carriers desired to alleviate some of these new high-cost risks. They did so by crafting an exclusion to liability coverage for claims for environmental contamination that the insured and its carrier did not expressly include in the policy terms.

A contemporary pollution exclusion clause reads as follows:

> This insurance does not apply: . . . (f) to bodily injury or property damage arising out of the discharge, dispersal, release or escape of smoke, vapors, soot, fumes, acids, alkalis, toxic chemicals, liquids or gases, waste materials or other irritants, contaminants or pollutants into or upon land, the atmosphere or any watercourse or body of water; but this exclusion does not apply if such discharge, dispersal, release or escape is sudden and accidental.[13]

20.2.2 Relationship to the "Products" and "Completed Operations" Coverage

The CGL policy describes the ''products'' hazard as ''bodily injury and property damage arising out of the named insured's products or reliance upon a representation or warranty made at any time thereto, but only if the bodily injury or property damage occurs away from the premises owned by or rented to the named insured and after physical possession of such products has been relinquished to others.'' Coverage for the ''products hazard'' must be specifically and separately purchased by the insured, as is also true of coverage for ''completed operations''. The ''completed operations'' hazard includes damage occuring *after* the insured has ''completed or abandoned a construction, installation, repair or servicing activity.''[14]

The court in *CPS Chemical Co. v. Continental Insurance Co.*[15] explained the relationship between the pollution exclusion and the ''products'' or ''completed operations'' exclusions. The court concluded that the pollution exclusion operates as a restatement of the definition of what is a compensable ''occurrence'' within the meaning of the CGL policy. In that suit, where a contractor engaged by the insured was claimed to have dumped hazardous wastes at a municipal dump site, the carrier claimed it had no duty to defend. The New Jersey trial court disagreed, reasoning that: (1) toxic waste disposal was an ''occurrence'' under the CGL policy and (2) neither the products nor the completed operations hazards risks were implicated as the suit did not involve products, service operations, or construction by the insured. The court added that its decision was a ''reaffirmation of the principle that coverage will not be provided for intended results of intentional acts, but will be provided for the unintended results of an intentional act.''

20.2.3 The Meaning of "Sudden and Accidental"

Because there are such great costs involved in environmental liability, the question of when does the exclusion or its sudden and accidental exception apply have been contested vigorously. While the phrase ''sudden and accidental'' is not defined, the meaning of these words is the most frequent cause for litigation. The insurers contend that an event that does not end quickly is not sudden and accidental while the insureds argue unintended events of which they had little or no notice are sudden and accidental regardless of duration. There is an apparent ambiguity in the meaning of sudden and accidental. On one hand they can have a temporal meaning and on the other hand they can be referring to an event which causes an unexpected result. The judiciary seems to be evenly split on the issue.[16]

While most courts have concluded that for the exception to apply the occurrence must be both sudden and accidental, some have held that the terms sudden and accidental merely restate the neither expected nor intended language from the CGL definition of occurrence. Other courts have defined sudden to mean a short period of time.

The federal appeals court for the Third Circuit, interpreting Pennsylvania law, addressed the meaning of "sudden and accidental" in *Northern Insurance Co. of New York v. Aardvark Assoc., Inc.*[17] In that suit, the two insurance companies were granted summary judgment against Aardvark Assoc., a hauler of industrial waste. In 1977 and 1978 Aardvark had transported industrial waste to two separate sites in Pennsylvania. The EPA became aware of the sites in 1981 and 1982, in 1983 the EPA notified Aardvark that it was a "potentially responsible party" under the Comprehensive Environmental Response, Compensation, and Liability Act (CERCLA). In 1985 the EPA filed suits against Aardvark to recover the clean-up cost of the sites. Aardvark notified its insurance companies of the situation shortly after receiving the initial EPA notices. Aardvark's insurance policies contained the conventional pollution exclusion clauses, including the exception for "sudden and accidental" discharges.

The parties filed claims and cross-claims for declaratory judgments regarding their obligations. The insurance companies asserted that the term "sudden and accidental" was clear and unambiguously required that the discharge be abrupt and short in duration, i.e., "sudden", and unintended, i.e., "accidental". Agreeing with the proposition that the phrase "sudden and accidental" was "clear and plain, something only a lawyer's ingenuity could make ambiguous," the court reasoned that even if the word sudden included unexpected, the term had to have an additional element because unexpected is included in accidental. The words would not be linked together, the court observed, if sudden did not include an additional meaning, which additional meaning is defined temporally and denotes abruptness and brevity. The court noted further that Aardvark's pollution was not sudden, as it was done gradually over a period of years with many individual shipments of waste.[18]

Similarly, the federal Court of Appeals for the Second Circuit interpreted the "sudden and accidental" exception to the pollution exclusion clause to have a plain meaning similar to the Third Circuit's definition. In *State of New York v. Amro Realty Co.*,[19] Amro, the alleged polluter, was charged by the state under CERCLA and public nuisance law of dumping hazardous waste into its drainage ditch groundwater via its septic tank, and directly into the ground for a period of approximately 30 years. First the court looked at the word accidental and determined that where the complaint alleges the discharge was intentional and deliberate it cannot be considered accidental. The key idea was that intentional acts are not accidental. Amro argued that they may have intentionally discharged the waste, but they did not intend to pollute as they simply released waste into their drainage system and did not know where it would end up. The insured tried to characterize its action as being negligent, hence it was unintentional thus accidental.

The court found that the characterization of negligence, when looked at in the context of the discharge, could not turn the discharge into an accident. The court noted that "'disposal' connotes a deliberate intentional activity."[20] The court also stated that purposeful conduct could not be considered accidental, and concluded that the discharge could not be considered accidental merely because the insured did not affirmatively know where the waste might migrate ultimately.[21]

The court then went on to evaluate the meaning of the term "sudden", and stated that even had they considered the discharge to be accidental, there still would be no coverage because the discharge was not sudden. The exception will only apply if the discharge is both sudden and accidental; if the discharge was neither sudden nor accidental the occurrence is not covered. The court read sudden to mean a release over a short period of time. As this discharge was continuous over approximately 30 years, it was not sudden.[22] The Court of Appeals for the Sixth Circuit has reached a congruent interpretation of the "sudden and accidental" clause.[23]

20.2.4 Pollution Exclusion Applicability to "Active" Polluters

In one suit, an alleged polluter whose insurers had tried to invoke the pollution exclusion clause argued that the clause only applied to "active" not "passive" polluters. The company was a transport company and claimed that it only "passively" polluted as it did not produce the waste, it merely transported it and discarded it at the dump site. The court looked at the language of the clause and pointed out that the clause makes no reference to either "active" or "passive" polluters therefore no distinction should be made under the law. It concluded that the terms "active" and "passive" were extraneous to the issue of the clause's application.[24]

20.2.5 Burden Of Proof

The federal Courts of Appeal for the Second and Third Circuits have reached different conclusions on the question of whether the insured or the insurer bears the burden of proving the applicability of the pollution exclusion clause. In *Aardvark Assoc., Inc.*,[25] the Third Circuit held that the insured party, in that case, had the burden of establishing that the claims against it fell within the "sudden and accidental" exception. The court noted that while it is generally understood that the insurer has the burden of proving an exception to coverage the courts are split on who must prove an exception to an exclusion.[26] Reaching the seemingly opposite conclusion, the Second Circuit held in *State of New York v. Amro* that "when an exclusion clause is relied upon to deny coverage, the insurer has the burden of demonstrating that the 'allegations of the complaint cast the pleading solely and entirely within the policy exclusions, and further, that the allegations, *in toto*, are subject to no other interpretation.'"[27]

20.2.6 Timely Notice Of Claim

Most insurance policies contain clauses allowing the insurer to deny coverage when it is not notified of a possible claim against it in a timely manner. Some claims must immediately be reported to the insurer while others must be reported as soon as practicable. The Second Circuit articulated a test for timely notice of occurrences in *Ogden Corporation v. The Travelers Indemnity Co.*[28] In *Ogden*, the insured was first informed by the company it leased property from that the lessor expected them to pay for the cost of environmental cleanup mandated by the state on July 25, 1985. The Insured sent a letter to its lessor on June 27, 1986 disclaiming liability for the environmental damage to the leased property. The insured did not notify its insurer of the possible occurrence claim until April 22, 1988. The court articulated that the test is whether the circumstances known to the insured at a particular time would have suggested to a reasonable person the possibility of a claim against the insurer, and found that the 2.5-year lapse from the time the insured first became aware of a possible claim and when they notified the insurer was unreasonable.

The insurer's right to deny defense or indemnification under a CGL policy because it did not receive timely notice is not absolute. The carrier must elect its option within a certain time. Where an insurer is notified of a claim and it is sufficiently late to permit it to invoke the timely notice exclusion, but the insurer fails to do so while denying coverage on other grounds, it waives the defense of untimely notice. For instance, if an insurance company after receiving untimely notice denies coverage for another cause but not untimely notice and that other cause is found to not be sufficient to deny coverage, the insurer cannot turn around and later plead untimely notice. An insurer's action to disclaim on certain grounds but not others is deemed to be conclusive evidence that the insurer intends to waive the unasserted grounds.[29]

20.3 COVERAGE FOR DUTY-TO-WARN CLAIMS

It is well established that a product manufacturer may be liable in tort or warranty for damages where it fails to provide adequate warnings about risks of use or exposure to its product. Liability may also be found where the manufacturer fails to provide adequate instructions as to how the product may be used without an unreasonable risk of physical injury or property damage. Conceptually, the absence of adequate warnings or instructions is often said to render the product defective.

The analogy between warning inadequacy and defective condition has a logic in primary actions against the manufacturer, for it provides the fact-finder with a logical deductive path for evaluating whether the product defect rendered it unreasonably dangerous, or unmerchantable, under state law. The proper relationship between warning inadequacy and product defect is not nearly so clear when interpreting carrier duties to products liability defendant manufacturers where the injured claimant states, as one of several alternative causes of action, that the seller's failure to provide sufficient warnings was a proximate cause of plaintiff's physical injury or property loss.

There is growing authority that a CGL policy's exclusion of coverage for the "products hazard" does not extinguish the carrier's duty to defend a product-related claim alleging the manufacturer's failure to provide adequate warnings.[30] Unlike products liability actions, where the inadequate warning-defective condition analogy is grounded in tort and products liability doctrine, the reciprocal duties of a carrier and its insured are defined by the contractual language of the policy. The gravamen of the carrier's duty to defend is whether the complaint states facts that could bring the case within the coverage of the policy. Thus a carrier may have obligations under a policy where the complaint alleges one or more claims that may be within the liability coverage, even if other, or even most, of plaintiff's claims are excluded from coverage. Although the general rule in insurance coverage declaratory judgment actions is that the court will try to interpret policy obligations according to their plain and ordinary meaning, there is measurable judicial sentiment that "overly subtle or technical interpretations may not be used to defeat reasonable expectations of insureds."[31]

The language of the CGL policy excludes coverage for the "products hazard" unless the insured secures it specifically. As referenced earlier, a conventional description of the "products hazard" includes "bodily injury and property damage arising out of the named insured's products or reliance upon a representation or warranty made with respect thereto," but does not expressly include liability for an insured's *omission* (failure) to warn or to provide adequate instructions.

The issue often arises as to whether an insured who has omitted to secure "products hazard" or "completed operations hazard" coverage may nonetheless demand that its carrier defend and indemnify it in claims alleging, among other counts, that it failed to warn or provide adequate instructions to a third party. How is coverage to be interpreted in a products liability claim where the product conforms to the ordinary description for like goods, and performs as expected, but where claimant alleges a failure to warn of risks unassociated with faulty design or fabrication?

There is increasing decisional authority holding that a failure-to-warn claim is in the nature of the general risk of doing business that would motivate an insured to purchase liability coverage and therefore will create a duty to defend even absent coverage for the "products hazard". A leading decision so holding is *Cooling v. United States Fidelity and Guaranty Corp.*[32] In *Cooling,* the Louisiana court reiterated the general rule that as policy language is drafted by the carriers, ambiguous language will be strictly construed against the insurer, and that a carrier "has the duty to clearly express exclusions to its obligations."

Carrier duties may be triggered by claims of failure to warn even where policy "products hazard" exclusionary language is quite particular. One example is *Scarborough v. Northern Assurance Co. of America,*[33] a coverage declaratory judgment action following a suit brought against the manufacturer and a sand supplier by an employee of a purchaser, a sandblasting concern. The employee, who contracted silicosis, claimed that defendant failed to warn about the risks of that disease and the means of avoiding it. The policy exclusions at issue there included the exclusion of liability for failure of the insured's product "due to a mistake or deficiency in any . . . printed instructions prepared or developed by any insured[.]" Notwithstanding the policy language seemingly excluding coverage for informational "deficienc[ies]", arguably a failure to act, the court ruled that plaintiff's general allegation of failure to warn, not limited to the adequacy or inadequacy of written instructions, was not excluded by the carrier's policy.

Other authority suggests that a failure-to-warn claim will raise carrier duties even without products coverage where, apart from the claimed lack of information for the user or consumer, the product is otherwise adequately designed and performs as intended. Although holding that the exclusion in question relieved the carrier of defense duties, some decisions have recognized that on other facts, claims of seller negligence, i.e., the failure to provide sufficient warnings or instructions, may, depending upon the facts, constitute "an act sufficiently removed from the quality of the product in question [to] escape the exclusionary clause."[34]

Most recently, a Pennsylvania trial court gave a congruent interpretation to the products hazard exclusion, finding that it did not preclude coverage of a failure to warn claim. In *Hartford Mutual Insurance Co. v. Moorhead,*[35] the policy specifically excluded "bodily injury or property damage included within the . . . products hazard," with the policy rider describing the "products hazard" as including "bodily injury or property damage arising out of the named insured's products or reliance upon a representation or warranty made at any time with respect thereto[.]"

Analogizing plaintiff's failure to warn claim to an allegation that defendant failed to exercise due care in performing a "service" following the manufacture and sale of the product, the trial court concluded that plaintiff's failure-to-warn claim was not within the products hazard exclusion. In so doing, the *Moorhead* court was careful to confine this rationale to failure-to-warn claims where the product, apart from the claimed informational deficiency, works or performs as expected. Where, on the other hand, a product failure is a producing cause of plaintiff's loss or injury, the court noted, decisions hold consistently that absence of products hazard coverage absolves carrier duties.[36]

NOTES

1. See generally Madden, "Products Liability Insurance Coverage and Exclusions," 10 *J. Prod. Liab.* 301 (1987).
2. Oshinsky, "Comprehensive General Liability Insurance and the Scope of Coverage in Long-Term Exposure Cases," 17 *Forum* 1035 (1982).
3. 924 F.2d 39 (2d Cir. 1991).
4. *Id.*, 924 F.2d at 41 (quoting New York authority).
5. *Id.*
6. 936 F.2d 1420 (2d Cir. 1991).
7. *Id.*, 936 F.2d at 1426–1427.
8. E.g., *J. H. France Refractories Co., v. Allstate Insurance Co.*, 578 A.2d 468 (Pa. Super. 1990).
9. 667 F.2d 1034 (D.C. Cir. 1981), *cert. denied* 455 U.S. 1007 (1982), rehearing denied 456 U.S. 951 (1982), on remand 597 F. Supp. 934 (D.D.C. 1984).
10. E.g., *J. H. France Refractories Co. v. Allstate Insurance Co.*, 578 A.2d 468 (Pa. Super. 1990) (entire asbestos disease process from first exposure through manifestation).
11. *Id.*, 578 A.2d 468 (Pa. Super. 1990).
12. *Dow Chemical Co. v. Associate Indemnity Corp.*, 727 F. Supp. 1524 (E.D. Mich. 1989).
13. See generally *USF&G v. George Whiteside Co.*, 937 F.2d 1169 (6th Cir. 1991).
14. Significantly, the CGL policy's interpretive language governing the completed operations hazard states that "[o]perations which may require further service or maintenance work, or correction, repair, or replacement because of any defect or deficiency, but which are otherwise complete, shall be deemed completed."
15. 489 A.2d 1265 (N.J. Law Div. 1984).
16. See *Northern Insurance Co. of New York v. Aardvark Assoc. Inc. and Insurance Co. of North America,* 942 F.2d 189 (3d Cir. 1991).
17. 942 F.2d 189 (3rd Cir. 1991).
18. *Id.*, 942 F.2d at 191, 192.
19. 936 F.2d 1420 (2d Cir. 1991).
20. *Id.*, 936 F.2d at 1428.
21. *Id.*, 936 F.2d at 1427, 1428.
22. *Id.*
23. See *United States Fidelity and Guaranty Co. v. George Whitesides Co., Inc.*, 932 F.2d 1169, 33 E.R.C. (BNA) 1409 (6th Cir. 1991).
24. *Northern Insurance Co. of New York v. Aardvark Assoc., Inc.*, supra note 16.

25. 942 F.2d 189 (3d Cir. 1991).
26. *Id.*, citing *Firemans Insurance Co. v. Ex-Cello-Corp.*, 702 F. Supp. 1317, 1328 (E.D. Mich. 1988).
27. 936 F.2d 1420, 1427 (2d Cir. 1991) (citing *International Paper Co. v. Continental Casualty Co.* 35 N.Y.2d 322, 325, 361 N.Y.S.2d 873 (1974).
28. Supra note 3, 924 F.2d at 42.
29. See *State of New York v. Amro,* 936 F.2d 1420, 1429–1433 (2d Cir. 1991).
30. E.g., *Hartford Mutual Ins. Co. v. Moorhead,* 396 Pa. Super. 234, 578 A.2d 492 (Pa. Super. 1990) (collecting authority). See generally Madden, The "Products" and "Completed Operations" Exclusions and the Allegation of Failure to Warn, 7 *J. Prod. Liab.* 205 (1984).
31. *Moorhead, id.*, 578 A.2d 492 (Pa. Super 1990).
32. 269 So.2d 294 (La. App. 1972).
33. 718 F.2d 130 (5th Cir. 1983).
34. E.g., *Cravens, Dargen & Co. v. Pacific Indemnity Co.*, 105 Cal. Rptr. 607 (1972).
35. 396 Pa. Super. 234, 578 A.2d 492 (Pa. Super. 1990).
36. E.g., *Viger v. Commercial Union Insurance Co.*, 707 F.2d 769 (3d Cir. 1983); *K. C. Mfg. Co. v. Shelby Mutual Insurance Co.*, 434 So.2d 1004 (Fla. 1983).

General Index

Case Index